THE CRITICS RELISH MAYONNAISE AND THE ORIGIN OF LIFE

"Witty and whimsical. Morowitz bestows his light touch on the driest of subjects."

— *Newsday*

"Harold Morowitz's prose is relaxed, elegant.... His book is exemplary...because the writing is so good and the author's interests are so wide."

— San Francisco *Chronicle*

"Wonderfully diverting and very wise."

— *Kirkus Reviews*

"Morowitz belongs to the band of doctors and biologists, including Stephen Jay Gould, Gerald Weissman, Lewis Thomas, etc., who are keeping the essay alive as a literary form.... Delightedly recommended."

— *Booklist*

"This brand of 'Mayonnaise' is an intellectual treat."

— *Southern Connecticut Newspapers, Inc.*

Most Berkley Books are available at special quantity discounts for bulk purchases for sales promotions, premiums, fund raising, or educational use. Special books or book excerpts can also be created to fit specific needs.

For details, write or telephone Special Markets, The Berkley Publishing Group, 200 Madison Avenue, New York, New York 10016; (212) 686-9820.

MAYONNAISE AND THE ORIGIN OF LIFE

THOUGHTS OF MINDS AND MOLECULES

HAROLD J. MOROWITZ

Harold J Morowitz

BERKLEY BOOKS, NEW YORK

All essays originally published in *Hospital Practice* except for the following: "Navels of Eden," "Beetles, Ecologists and Flies," "Biochemistry is Beautiful," "Ice on the Rocks," "The Beauty of Mathematics" and "Facts and Artifacts," originally published in *Science 82*; "Two Views of Life," originally published in *Science 83*; "Rediscovering the Mind" and "Do Bacteria Think?" originally published in *Psychology Today*; all reprinted by permission. "Reducing Life to Physics," © 1980 by The New York Times Company. Reprinted by permission.

This Berkley book contains the complete
text of the original hardcover edition.

MAYONNAISE AND THE ORIGIN OF LIFE
THOUGHTS OF MINDS AND MOLECULES

A Berkley Book / published by arrangement with
Charles Scribner's Sons

PRINTING HISTORY
Scribner's edition published 1985
Berkley edition / January 1986

ISBN: 0-425-09566-5

A BERKLEY BOOK ® TM 757,375
Berkley Books are published by The Berkley Publishing Group,
200 Madison Avenue, New York, NY 10016.
The name "BERKLEY" and the stylized "B" with design
are trademarks belonging to Berkley Publishing Corporation.

PRINTED IN THE UNITED STATES OF AMERICA

Contents

SECTION ONE

THE GALÁPAGOS

Some of Our Trials about Evolution

The Galápagos 3
Navels of Eden 7
Tell It to the Judge 11
When the Chips Are Down 16
The Phenomenon of Man 20
Mayonnaise and the Origin of Life 27
Beetles, Ecologists, and Flies 31
A Leap of the Imagination 36
Two Views of Life 40
Bird What? 45

SECTION TWO

THE ROOTS OF WISDOM

Thoughts about Medicine and Dentistry

The Roots of Wisdom 51
My Mouth Shall Praise 55
Between Gargoylism and Gas Gangrene 59
Jurisgenic Disease 64
Geriatric Gobbledygook 68

Trees and Forests 72
Much Ado about Nothing 76
Through an Ophthalmoscope Darkly 80
Little Black Boxes 84
The Ecosystem Within 89

SECTION THREE
THE PACE OF LIFE
Mind over Matter

The Pace of Life 95
The Time of Your Life 99
Omar and the Ayatollahs 104
The Paradox of Paradoxes 108
St. James Infirmary Blues 113
Do Bacteria Think? 117
ESP and dQ over T 122
The Hit Parade 126

SECTION FOUR
PRISON OF SOCRATES
Some Social Issues

Prison of Socrates 133
Turning Colorado into Kansas 137
Welcome, Class of 1984 141
The Infernal Combustion Engine 146
Legacy 150
Facts and Artifacts 154

SECTION FIVE
DEALING FROM A FULL DECK
Unusual Individuals and Their Work

Dealing from a Full Deck 161
As the World Turns 166
Gossamer Birds, Flights of the Imagination 170
Dateline: Dubrovnik 174
On Making a Point 178

Before Him Lay a Sea of Lies 182
Warm-Blooded Fish? 186
How E. coli *Got Its Name* 190

SECTION SIX
BIOCHEMISTRY IS BEAUTIFUL
Science and Esthetics

Biochemistry Is Beautiful 197
Ice on the Rocks 201
Biphenyls and Bipeds 205
Living Lodestones 209
From Soup to Solid-State 214
Entropy Anyone? 218
The Odd Couple 223
Energy Flow in Ulysses 227
The Beauty of Mathematics 232

Index 237

SECTION ONE

THE GALÁPAGOS

Some of Our Trials about Evolution

The Galápagos

For a biologist a trip to the Galápagos is a professional pilgrimage. If it lacks the religious intensity of the road to Mecca, it perhaps resembles the good-natured journey to Canterbury depicted by Chaucer. Whatever our specialty in the life sciences, our historical roots lie here among the lava rocks. We are seeing the same sights that greeted our intellectual forefather, and the existential quality of that experience is enough. A simple walk through a thousand nesting boobies may not be a religious experience, but it has a deep spiritual quality. Looking an iguana in the eye at three feet conveys its own sense of awe, and maybe that too is what pilgrimages are all about.

Having arrived at the shrine, the wondrous animals, strange plants, and awesome physical surroundings evoke unusual thoughts. Standing atop a jagged lava peak and gazing out over portions of the archipelago, one can fantasize about the nature of our literary heritage if Dante had visited here. Paradise, perchance, would have emerged with man and animals enjoying a mutual trust as painted by Edward Hicks in *The Peaceable Kingdom* and extolled by Isaiah. Inferno might have been boldly drawn with endless lava fields, sharp rocks, and ominous cinder cones. A still active volcano or two would have warmed the image even further.

However, these islands became known to Western thought not through the fine Italian mind of Dante but by way of the usually factual prose of Charles Robert Darwin. Chapter XVII of *The Voyage of the Beagle* describes the flora and fauna and conveys the naturalist's sense of wonderment. The usually placid traveler was almost moved to poetry when prophetically he wrote:

> Considering the small size of these islands, we feel the more astonished at the number of their aboriginal beings, and at their confused range. Seeing every height crowned with its crater, and the boundaries of most of the lava streams still distinct, we are led to believe that within a period, geologically recent, the unbroken ocean was here spread out. Hence, both in space and time, we seem to be brought somewhat nearer to that great fact—that mystery of mysteries—the first appearance of new beings on this earth.

Some 20 years later, when Darwin completed *The Origin of Species*, he was deeply influenced by what he had seen and collected in the Galápagos. These islands were rich with endemic species that occurred nowhere else in the world yet bore striking affinities to the inhabitants of the South American mainland some 500 miles to the east. He reasoned that the plants and animals had arrived as immigrants and had, in each isolated locale, been modified by natural selection, with favorable variations being decided by the existing conditions of the habitat. The Galápagos, as seen by Darwin, were neither heaven nor hell but the isles of purgatory where each species was tested. The fate of each was held in the balance, with survival, change, or extinction as the possible outcome. So it is with all evolution; only extinction is definitely decided for all time, and each form of life must continually face the trial.

Thoughts of Darwin were never far from our minds as we followed the trail seaward on the island of Santa Fé (Darwin called it Barrington, but then Captain Fitz Roy doubtless permitted only the use of English names). We were somewhat tired from the morning climb but also elated at having seen a land iguana and a Galápagos hawk. About 300 meters from the beach a sea lion pup emerged from among the crags onto the trail. Its behavior was different from that of the hundreds of juveniles we had watched on the rocks and sands of these islands. Instead of cautiously keeping its distance, this sea lion was constantly coming toward us, trying to nuzzle up to our feet. We had been warned not to touch any of the animals or to impart any odors that might interfere with maternal recognition, for which smell plays a major role. We kept retreating from this creature who was soliciting our contact.

The pup was about six weeks old and emaciated as compared with members of its cohort. It followed us to the beach, where about 30 mature females and juveniles were lying about in the sun, while a large bull patrolled the waters just offshore. The underfed infant went from sea lion to sea lion, trying to establish contact. The other pups tried playing with it, but the mature females showed rejection either by jumping away or driving the stray off, sometimes with a nip. When the pathetic beast wandered into the water, the bull escorted it back to shore. There was no protection from the starvation that was leading this frantic young animal from adult to adult looking for an active mammary gland.

We were witnessing a youngster separated from its mother. She may have abandoned it for unknown reasons, she may have been injured or killed, or the two simply may have lost each other on that complicated shoreline. Whatever the reason, no other female would feed the stray. There is a strong maternal attachment among these mammals, but each mother is totally committed to one recognized baby and to no other. Social laws are very strong among the pinnipeds. The survival of her own offspring leaves no nursing female the option of adopting an abandoned waif.

From a biological point of view the outcome was very clear. That pup was going to die of starvation; the laws of nature were inexorably grinding away. Either the young animal or its parent had somehow failed, and those genes for failure were being eliminated. The species was becoming better adapted to its life on Santa Fé Island. The scientific description pales before the intensity of the event. An evolutionary scenario was being played before us, and the drama was interrupted by the arrival of our skiff. As scientists, we would have liked to remain and observe the full behavioral repertoire unfold. As persons of feeling, we were glad to be relieved of participation in a cruel scene. If these are the isles of purgatory, there is also a bit of hell in not being able to help a suffering creature pleading to you with its soulful eyes. In violation of park regulations against leaving any biological materials behind, we may have dropped a tear or two on Santa Fé.

Although our experience had its gloomy aspects, these islands can also induce a sense of joy. There is wonderment and beauty and a thousand sights and sounds to stimulate the mind and soothe the spirit. There is a warm sun, cold sea current, and a collection of birds like no place else on earth. And I am sure that is what an earthly purgatory must be like—having some of both extremes.

Our pilgrimage to the islands of Darwin has brought us face to face with the master's lessons about survival and change. He enunciated profound principles that echo through all of evolution and ecology. Still, it is one thing to deal with natural selection as an abstract textbook concept and quite another to stand on a lonely beach and helplessly watch a tiny sea lion starve to death.

Navels of Eden

With shades of Clarence Darrow and William Jennings Bryan in the background, the "monkey trial" appears to be resuming as a regular event in the American system of jurisprudence. The dispute over a biology instructor's freedom to teach his discipline has reappeared with some surprising twists. The fundamentalists now argue through their own scientists, a small but loyal group of creationists who try to formulate a Genesis-centered version of the origins of the universe, the planet, and the biota. Rather than rejecting science outright, they maintain that the first book of the Bible represents the best of geophysical and biological truths. These creationists, speaking in the language of thermodynamics and biochemistry, have exerted a powerful influence on state legislatures and boards of education.

The efforts of this contemporary group of theologically oriented scientists oddly recall the work of their most distinguished intellectual forerunner, a man who tried to refute Darwin's *The Origin of Species* a full two years before that book was written. Philip Henry Gosse was an experimental zoologist and popular author of great distinction who spent his Sunday mornings as a lay preacher for the Plymouth Brethren, a Protestant sect holding extreme fundamentalist views. He thus combined within one individual the warring factions of 19th-century English thought. His conflicts and travails illustrate some profoundly human aspects of the science-and-religion saga.

In the 1850s Gosse was at the height of his career. He was an acquaintance of Professor Thomas Huxley and for a while a correspondent of Dr. Wilberforce, Bishop of Oxford. Gosse was elected to the Royal Society on the basis of his classical work on rotifers, a group of small primitive animals. In 1855 he was introduced to Charles Darwin at a meeting of the Linnaean Society. The letters between them began in 1856, with Darwin stating, "I am working hard at the general question of variation and paying for this end special attention to pigeons." The founder of evolutionary theory was collecting data, and naturalist Gosse supplied a variety of facts on baldpate pigeons, canaries, transport of plants and animals to distant islands, and fighting among male crustaceans.

It was clear to biologists and geologists alike that "evolution" was in the air. A wealth of paleontological and physical data made it obvious that the earth was far older than Bishop Ussher's chronology of some 6,000 years. The biological facts were also piling up, while at Down House the cautious Charles Darwin continued to amass data supporting his unpublished views. For Philip Gosse it was a difficult time. Gosse the scientist was convinced of the geological data, while Gosse the preacher was absolutely committed to the Word of God. His son states, "He was a visionary on one side of the brain though a biologist on the other." Personal woes added to his ideological dilemma with the fatal illness of his wife Emily, his companion and religious coworker. In his melancholy brooding he set out to resolve the terrible conflict that enveloped him.

Out of the depths he came forth with the book *Omphalos, an Attempt to Untie the Geological Knot*. Gosse argued that since all organic nature moves in a circle, creation must occur at some point in the cycle as a violent interruption and must bear the evidence of the prior processes as if they had existed. He therefore maintained that creation burst forth with false records of a nonexistent past. The word *prochronic* was coined to describe such objects. Thus he assumed that the Creator had made trees in the garden of Eden with annual rings for years that had never passed and had created Adam with a navel, notwithstanding the fact that the first man had never existed *in utero*. Although it is hard to grasp from our present perspective, the issue of whether there were navels in Eden was a serious intellectual problem for many highly intelligent mid-19th-century scholars. The book's title, *Omphalos*, is simply the Greek word for umbilicus. Whimsical ideas abounded because the classical account of creation and a 6,000-year-old earth had become inconsistent with the evidence from many disciplines, and the faithful were desperate. Gosse struggled passionately to defend his faith, and his work "was received with scorn by the world of science and with neglect by the general public." Indeed the book was rejected by both sides with equal vigor. Gosse's friend and fellow believer Charles Kingsley wrote to tell him, "In the case of fossils which *pretend* to be the bones of dead animals and your newly created Adam's navel, you make God tell a lie. . . . I cannot give up the painful and slow conclusion of five and twenty years of study of geology, and believe that God has written on the rocks one enormous and superfluous lie for all mankind."

In 1859 *The Origin of Species* appeared in print, and *Omphalos*, along with hundreds of other books, was consigned to the back shelves of library stacks. Evolution won the day among practicing scientists, and the celebrated battles between natural historians and theologians began to be waged with great bitterness. Gosse moved from London to Devonshire, where he lost himself in the study of marine invertebrates and his ever ongoing work with the Plymouth Brethren "in the cottages of the sick and poor of his congregation." He withdrew from public controversy.

In 1863 the correspondence between Gosse and Darwin was revived. They were both orchid fanciers and exchanged advice on how to artificially fertilize some favorite plants. Their letters contain great quantities of detail on pollen masses, ovules, nectars, stigmatic cavities, and flowers. Not a word passes about evolution or creation or the enormous ideological gulf that separated the two great naturalists. The letters are quaint and polite and very British. Both men were recluses; both could lose themselves in the study of rotifers or barnacles or in cultivating orchids. Both had been trained as clergymen. Darwin gave the world one of its greatest books, while Gosse's volume is all but lost.

The case of Philip Gosse is a poignant example of an individual who was devoted to science yet brought to his subject an inflexible prior commitment to an absolute truth. This left him in the position of holding contradictory views on a number of subjects. Since logicians have long known that two contradictory premises can lead to any conclusion whatsoever, it is not unexpected that Philip Gosse produced a very enigmatic book. The reason for wiping the dust off this seldom read volume is to remind present-day fundamentalist scientists of the difficulties of assuming that reading an English translation of a very old Hebrew book takes precedence over reading much older rocks if we wish to establish the most accurate possible history of our planet and the life thereon. In a free society one can, of course, accept contradictions and conclude anything. That procedure has never led to particularly good science—or good theology either, for that matter. Our sympathy for the pathos of Philip Gosse the man cannot extend to a sympathy for a scientist's gross intellectual errors about belly buttons in Paradise.

Tell It to the Judge

As the airplane circled over the rice paddies between Agra and New Delhi, there came to mind a book on the theory of relativity that I had read many years ago. That work used the phrase *world-line* to describe an object's path in the four dimensions of space and time. Through the mental haze of jet lag I was acutely conscious of how strange my world-line had been for the past few days. In space my itinerary had gone from New Haven to New Delhi by way of Little Rock. The time had been compressed into less than a week. Not too many hours ago I had been in a plane circling over the rice fields of Arkansas. The common theme of growing grain for food was at least one tangible feature in an otherwise abstract set of circumstances.

The first leg of my journey to the East had led westward to the Federal District Court in Little Rock, presided over by Judge William Overton. The large courtroom was crowded with reporters and spectators assembled to see and hear *McLean* v. *Arkansas*. The court case had its origins in, or was created by, Act 590 of 1981, passed by the Arkansas legislature and duly signed on March 3, 1981, by Governor Frank White. The legislation mandated "Balanced Treatment of Creation-Science and

Evolution-Science in the Public Schools." It had been challenged by Reverend McLean, the American Civil Liberties Union, and a number of other individuals and organizations. The basis of the challenge asserted that since there exists no scientific discipline of creation-science, such a term was being used as a cover to require teaching the Book of Genesis in public schools. This requirement, according to the plaintiffs, was in violation of that part of the First Amendment that states, "Congress shall make no law respecting an establishment of religion." The trial began on December 7, 1981.

Creation-science is defined in the legislation as "the scientific evidences and related inferences that indicate:

1. Sudden creation of the universe, energy, and life from nothing;
2. The insufficiency of mutation and natural selection in bringing about development of all living kinds from a single organism;
3. Changes only within fixed limits of originally created kinds of plants and animals;
4. Separate ancestry for man and apes;
5. Explanation of the earth's geology by catastrophism, including the occurring of a worldwide flood;
6. A relatively recent inception of the earth and living kinds."

In the plaintiff's presentation, the diverse parts of this definition hang together only insofar as they all come from the Book of Genesis. Indeed, the use of the word *kinds* instead of *species* echoes the King James translation of the Bible.

My presence in Judge Overton's courtroom related to the aspect of so-called evolution-science dealing with "the emergence of life from nonlife." The creationists argue that the second law of thermodynamics precludes biogenesis by natural processes; therefore supernatural events were required. The greater part of my career has been devoted to the thermodynamic foundations of biology and biological self-organization. This has been a very satisfying, if somewhat obscure, branch of research. Suddenly it was being brought to center stage. Life and the laws of thermal physics had become an issue of national attention. It was a strange thought to contemplate from the ivory-tower perspective, yet in the quiet judicial atmosphere of the courtroom the thermodynamic theory was a piece of evidence to be weighed and judged along with other pieces of evidence.

The celebrated second law of thermodynamics states that isolated systems move toward the maximum degree of molecular disorder. Isolation indicates the absence of flows of both matter and energy in and out of the system under consideration. The second law and the theory of evolution had been formulated in the same decade (1850–1860). At first there was great confusion within science itself, since physics seemed to be saying that, in time, things inevitably become disordered, while evolutionary biology was an example of increasing order. Some biologists maintained that living systems violated the laws of thermodynamics. Others maintained that life was a miracle, not subject to explanation by science.

In 1886 Ludwig Boltzmann, the noted Austrian physicist, resolved the confusion by pointing out that the earth is an open system undergoing a flow of solar energy. In such a situation the surface of the planet is not limited by a law that is restricted to isolated entities. Curiously, it took 60 years for Boltzmann's understanding to permeate the biological community, and some haven't gotten the message yet. The modern era of irreversible thermodynamics, ushered in by Lars Onsager in the 1930s, has gone further than Boltzmann by showing that systems inevitably become ordered under a flow of energy. Life happily involved no contradictions to the laws of physics.

The creation-scientists, wishing to preserve the need for miracles, have seized on parts of the argument of the past 130 years, attempting to prove that biological ordering violates the laws of physics and arguments from probability theory. It was unusual material to be telling the judge, but this strange case rested on many specialized branches of science.

It was necessary for me to leave Little Rock before the case was concluded. Nevertheless, I had a good feeling about the quiet, deliberative, judicial atmosphere of the court. Whichever way the case was decided, due process was being well served. Judge Overton ran a tight ship, yet everyone had his say.

Arkansas seemed far behind as the 747 set down in New Delhi, even though I was headed for the International Seminar on the Living State, a meeting to discuss our concepts of life, topics curiously and closely related to the legal procedures that I had just witnessed. The participants were physicists, chemists, biologists, and philosophers sharing a common concern with how the remarkable property we call life can arise from a collection of highly ordered atoms largely composed of carbon, hydrogen, nitrogen, oxygen, phosphorus, and sulfur. Physicists are getting more mystical these days, and a good deal of Eastern philosophy dealing with a world-mind was interwoven with some very sophisticated mathematics, thermodynamics, and statistical arguments. The chemically oriented among us argued for the great explanatory power of atomic theory in understanding living cells. Those concerned with the emergence of the mind faced more difficult problems.

As the conference drew to a close after six days of intense discussion I noticed on a table in the lobby outside the seminar room a set of pamphlets being distributed by the Bhaktivedanta Institute. To my surprise one tract dealt with the "Demonstration That Life Cannot Arise from Matter." Curiously enough, the Vedantic mystics are presenting some of the same arguments that led the creation-scientists to Little Rock. They, too, argue against the possibility of a scientific explanation of the emergence of life from inorganic matter.

I have traveled a full circle from the Bible-centered fundamentalists of Arkansas to the Krishna Consciousness followers of Swami Prabhupada. It is a curious meeting of East and West, joining in their mutual denial that the natural sciences can ever probe our origins. I am experiencing a strong desire to get back to the laboratory and those experiments that will give more detailed information on how energy flow leads to chemical ordering and life. Lines of Robert Frost come to mind:

> *The woods are lovely, dark and deep.*
> *But I have promises to keep.*
> *And miles to go before I sleep.*

When the Chips Are Down

Although the pursuit of science is ideally supposed to require innovation and creativity, one gets the feeling nowadays that grantsmanship, the pursuit of support, sometimes demands more real ingenuity than the search for nature's secrets. This anomaly, which has often bedeviled my own research efforts, was brought into focus by a recent visit to the biology department of a good, small college. About half of the faculty was without federal grants and expressed considerable disappointment. The current system of research-funding creates a class society. The haves possess access to research tools, assistants, and the route to success through frequent publication. They attend national meetings and fraternize with colleagues. They phone around the country to discuss the latest results. The have-nots, without any of these things, must struggle to do research as best they can with the equipment and supplies available around the department. It is clearly a situation where the rich get richer and the poor get poorer.

A second result of the reduced funding is the tendency of the haves to pick a safe path to avoid becoming have-nots. Researchers do not change fields but remain in those areas where they are known by peers sitting in judgment. In general there is a tendency to shun the highly innovative in favor of the more trodden paths. This is not an attack on peer review; rather it is a reminder that the system must often render judgments where no sharp assessments are possible. Elements of chance are prominent in such situations. I do not have a solution to this two-tier system in contemporary science education, although I believe one could be found easily and fairly cheaply, if we would only recognize the problem, by being less lavish with the haves and more generous with the have-nots. Rather than attempting to preempt the National Science Foundation by outlining all the details of such a plan, let us look at a unique solution to the funding problem offered by one of the great have-nots in the history of American science.

John Bell Hatcher was born in 1861. During a sickly childhood he was educated at home by his father, who was a part-time teacher. As an adolescent he was a miner in Iowa, where he became interested in the fossils associated with the coal beds. Using the savings from his labors, he enrolled in Yale College, where he came under the influence of the controversial paleontologist O. C. Marsh. Upon graduation the former student was hired by his professor to do collecting in the fossil beds of Kansas, which are rich in the bones of land turtles, mastodons, and rhinoceroses. After nine years of working many sites, during which he made extraordinary contributions to the Yale museum collection, Hatcher accepted an appointment as curator of vertebrate paleontology at Princeton.

Hatcher developed an interest in the extinct fossil vertebrates of Patagonia, which had been first described by Darwin in *The Voyage of the Beagle* and had subsequently been virtually ignored for 50 years. For two years he worked at Princeton, envisioning a collecting expedition to Argentina while spending time and effort privately raising money to support such an endeavor. From March 1896 to the late summer of 1899 Hatcher spent all but four months in the Southern Hemisphere on three separate collecting trips. The last of these was entirely supported by the paleontologist out of his own pocket. The support problem can perhaps best be seen in his own modest words in the acknowledgment section of the published work. "To the Lamport and Holt and the W. R. Grace S.S. Lines we were indebted for reduced rates of passage to and from South America." He had, it seems, virtually hitchhiked to his work site.

The scientist and his assistant, always short on money, worked under the most trying conditions. During one five-month period they did not see another human being. However, in periods when they did encounter other people, Hatcher executed one of the most daring plans for financial support of research that has ever been carried out.

As the expedition moved through Patagonia, this small, hard-bitten scientist out of the American West taught the local residents the game of poker. He was a shrewd player, and there was a flow of money from the gauchos and estancieros to the gringo. Hatcher's NSF, NIH, and Rockefeller Foundation all hung on the draw of a card as he worked his way through Argentina, flushing out support for his scientific endeavors.

James Terry Duce has written a fascinating account of John Bell Hatcher's last night in San Julian (*Atlantic Monthly*, September 1937). Poker players from miles around came to the port city intent on getting even. The evening's activities were described by Duce "as one of those friendly Western games with everybody's six-shooter on the table." As the play went on, the stacks of pesos in front of Hatcher began to grow into what a paleontologist would call a substantial deposit. At the appropriately dramatic moment, the steamer whistle blew, announcing Hatcher's departure. Some participants suggested that it was improper for him to leave at that point in the game. He rose, picked up his gun and his pesos, and backed through the door saying, "Goodnight, gentlemen." No one moved, and another grant had been made for the support of research.

After these trips Hatcher left Princeton and became curator of paleontology and osteology at the Carnegie Museum of Pittsburgh. Six years were spent by Princeton scientists in cataloging, classifying, and describing the enormous amount of material that had been shipped from Patagonia to New Jersey.

Ironically, the work of three impoverished years in the field was published in eight sumptuous volumes paid for by "the generosity of J. Pierpont Morgan, Esq., who has rendered it possible to publish these reports altogether worthy of their subject and with adequate illustrations." And so Princeton gave thanks, noting, "Mr. Morgan's liberality alone has put in one uniform series all the great and varied results of Mr. Hatcher's labors in South America."

In 1904 Hatcher contracted a fatal case of typhoid. In a life of only 43 years he had been instrumental in building up the collections of the Yale, Princeton, and Carnegie museums. Three full houses of fossils stand as a memorial to the intensity of his dedication to paleontological research.

Contemporary science has moved a long way from the cart and three horses that took Hatcher and his assistant across southern Argentina. An electron microscope, a scintillation counter, or time on a computer are not likely candidates for poker stakes. Auditing procedures force us into very directed ways of dealing with funding. Yet the spirit of John Bell Hatcher might convey a message to those struggling to get their research done. If you believe in yourself and what you are doing, don't stop! Life is cosmically unfair, but determination still helps. To the granting agencies, we point out a moral: far better would have been the support of Hatcher's struggles in Patagonia than the ex post facto beneficence of J. Pierpont Morgan to arrange fine packaging for those triumphs won in the dirt and sweat of the field.

The Phenomenon of Man

The period from the mid-1840s until the mid-1940s was characterized by a hundred years of warfare between evolutionary biologists and the religious establishment. A curious figure in that struggle was Pierre Teilhard de Chardin, a man who attempted the grand synthesis of opposing viewpoints. Born just a year before the death of Charles Darwin, young Pierre devoted himself with parallel enthusiasm to geological studies and entry into the Jesuit order. He matured into a paleontologist and a philosopher whose scientific investigation led him to a mystical view of the universal evolution toward a divine goal identified in some unclear way with the God of his Jesuit fathers.

Teilhard's major philosophical work, *Le Phénomène Humain* (*The Phenomenon of Man*), was completed in 1940 while he was in China studying the fossil remains of Peking man. His religious superiors denied permission for release of the work, and he remained true to his vows of obedience during repeated appeals to Rome. Most readers only became aware of his views in 1955, when friends arranged the posthumous publication of his manuscripts. The subsequent fate of his ideas has been extremely mixed, in part because of his ornate and often obscure literary style, which allows for a wide variety of interpretations. Nevertheless, his writings have caused enough furor in both science and theology that it may have taken these 30 years to approach a quiet assessment.

To many biologists still imbued with the early Darwinian fervor of Thomas Huxley and Ernst Haeckel, Teilhard was seen as renouncing causality, which they regarded as the true basis of science. Trained in a tradition that had been battling ecclesiastics for a century, evolutionists profoundly mistrusted the priest's credentials. They saw in his writings the despised doctrine of teleology, which states that the future is determined by some final goal rather than by some preexisting state of the system. The theologians, on the other hand, tended to see Teilhard de Chardin as an outsider putting forward a mystical deviation that replaced the sure orthodoxy of revelation with potentially heretical views that religious belief could be based on science. The Vatican issued a warning against uncritical acceptance of his ideas. Vehement attacks on his views have come from molecular biologists. In particular, his writings have been ridiculed by Jacques Monod and excoriated by Peter Medawar. In spite of all the attacks a Teilhard cult persists, and his influence is widely felt in the scholarly world.

In reexamining *The Phenomenon of Man* we shall try to cut through the "poetic grandeur" and doctrinal issues and see what part of the scientific argument remains today. The author himself wrote, "If this book is to be properly understood it must be read purely and simply as a scientific treatise." Few would agree with this assessment. Indeed, the principal criticism by biologists is that accepted theory, speculation, and metaphysics are so jumbled that it is impossible to sort them out. But if we are patient enough, there are ideas within the mixture that may be worthy of further scientific thought.

Teilhard argues that unfolding of cosmic history is a continuing natural evolutionary process marked by a certain few transition points when radical changes take place. He takes as an analogy the slow heating of water that gradually warms up to the boiling point, at which there is a discrete alteration of state. Continued heating will lead to a new change of temperature, but we are dealing no longer with liquid water but with steam, a gas with properties totally different from those of the liquid. So he views the major transitions: the appearance of matter, the formation of the earth, the origin of cells, and the rise of reflective thought.

What characterizes the major transitions is that they alter the universe so as to change completely the rules that are applicable. It is not that earlier laws are no longer valid but that new processes emerge and take preeminence in the evolutionary pattern. Teilhard primarily argues these changes from a scientific viewpoint, but they are obviously very important to him from a theological perspective that demands a separation of man from the rest of the living world.

The statements on the beginning of matter, although those of a "naturalist rather than a physicist," preview the big bang theory of modern astrophysics. Current exposition of these ideas in *God and the Astronomers* by Robert Jastrow and *The First Three Minutes* by Steven Weinberg comes close to the general views of Teilhard, who dramatically speaks of "an explosion pulverizing a primitive quasi-atom . . . then a swarming of elementary corpuscles." However, he characteristically insists on giving the event further meaning when he writes, "In its own way matter has obeyed from the beginning that great law of biology—'complexification.'"

Moving quickly through the formation of stars and planets and the geophysical history of the earth, Teilhard pauses at the next great transition, the advent of life. The systematic synthesis of small molecules, then of macromolecules, and finally of cellular structures is looked at as a natural progression obeying the laws of physics and chemistry. When the final emergence of life comes, he envisions it as a crisis that so changes the planet as almost to erase the traces of its prior chemical state. He thus refers to the "explosion of internal energy consequent upon and proportional to a fundamental super-organization of matter." Life is seen as an emergent property of atomic structures, but one so qualitatively different that its advent represents a cataclysmic event. Here again, most contemporary workers could accept the general outlines.

For the next few chapters the paleontologist unfolds for his readers the pattern of the tree of life from earliest unicellular forms to mammals. He sees it not as random but as directed and leading to the next major transition, the birth of reflective thought or the origin of man, which are regarded as overlapping. We need not look for an instantaneous start of reflective thought; it, too, evolved in a regular way among the hominids, but on a paleontologist's time scale this is a relatively rapid event. In Teilhard's words, "Discontinuity in continuity: that is how . . . the birth of thought, like that of life, presents itself." He coins a new word, *noogenesis*, to define the appearance of reflective thought, which is regarded as as profound an event as the origin of life itself.

It is precisely at this point that *The Phenomenon of Man* comes into deep conflict with much traditional biological thought. In *The Descent of Man* by Darwin and a number of related works a great effort was made to demonstrate that we have evolved from the primates. Since this was the key issue of the public battle over evolution, it was necessary to stress the closeness of modern man to his simian ancestors and to underplay the differences. This tendency has continued in biology right up until the present as a kind of overreaction to Bishop Wilberforce's taunting Huxley with the question of whether it was through his grandfather or grandmother that he claimed descent from a monkey.

In this setting the philosopher-priest entered the arena with the claim that the beginnings of reflective thought, paralleling the ascent of man, are as significant and as different a sequence of events as biogenesis. The scholar, who had devoted his professional life to studying fossils of man's immediate ancestors, was deeply aware of our animal nature and never attempted to minimize that fact. Yet he saw in the transition to thought a happening as explosive as any we can reconstruct and one that will accordingly change our planet in new and unforeseen ways.

The relevance of Teilhard de Chardin's thinking to contemporary biology now comes into focus. For if he is even partially correct about the radical nature of noogenesis, then the human sociobiologists are dead wrong in their assertion that the study of animal societies is critical to understanding human behavior. For those scientists the transition from animal to man involves no discontinuity. As an example, R. L. Trivers maintains: "The chimpanzee and human share about 99.5 per cent of their evolutionary history. . . . There exists no objective basis on which to elevate one species above another. . . . It is natural selection we must understand if we are to comprehend our identities." Thus human sociobiologists claim that the methods of animal ethology are appropriate to man. The confrontation between Teilhard's views and sociobiology is sharp and direct.

Those committed to zoological behavior studies as a basis of human understanding argue that the Jesuit scientist stressed discontinuity of reflective thought for theological reasons. He had to separate man from the beasts if he was going to bring him closer to the angels. On the other hand, it can be argued that the human sociobiologists are equally metaphysical in their approach. They have to argue that something radically new did not occur when reflective thought appeared on the scene. They have to bring man nearer to the beasts in order to deny the existence of anything but beasts, thus giving more meaning and importance to their animal studies. We are looking at the confrontation of two philosophical viewpoints, each rooted in science.

On the issue of the noosphere, Teilhard had no doubt that he was being scientific and empirical. It was obvious to him that the discontinuity was real. Referring to man, he asked, "Yet to judge by the biological results of his advent, is he not something altogether different?" He felt that his contemporaries had missed the obvious and neglected the essential factor that "when for the first time in a living creature instinct perceived itself in its own mirror, the whole world took a pace forward."

At this point, support for Teilhard's view of the crucial nature of the transition to thought comes unexpectedly from within the biological community. In a very perceptive essay called *Is History a Consequence of Evolution?*, L. B. Slobodkin moves the issue back toward science when he writes,

> I will develop the position that human history differs profoundly from the history of other organisms and differs to such an extent that the usual modes of evolutionary thought, in which organisms are "judged" in terms of their selective value, are simply not applicable.
>
> This argument will be based on biological properties of humans and some of the higher apes. In particular, I will claim that the biological capacity for the development of an environmentally determined self-image makes it impossible for human history, both public and private, to be predicted from considerations of evolutionary theory in any practically significant way.

Slobodkin then marshals the scientific arguments to support his view and brings to bear a number of experimental studies.

To argue that reflective thought is new is not to deny its biological roots or to separate it from physical events taking place in the central nervous system. Reflective thought is an emergent property of the cerebral circuits and exhibits novel features not previously seen. To use a computer analogy, the same hardware is employed, but a new software program makes possible operations not previously envisioned. Similarly, for Teilhard, thought grew out of the function of neural networks, but at some point a real discontinuity in the programming occurred. To assume that reflective thought is or is not a discontinuity must at the moment go somewhat beyond contemporary science, but on this issue, it is unfair to brand Teilhard as unscientific and accept human sociobiology as scientific.

The issue is clearly joined. Is the phenomenon of reflective thought a slight modification of the behavior pattern of lower organisms, or does it represent something sufficiently new to force us to change the rules by which we study human activities? Is reflective thought continuous in natural selection and normal biological processes, or is it a radically new phenomenon that must be faced on its own terms? At the moment it is a scientific question that cannot be completely dissociated from the metaphysical view we bring to our analysis.

It is now 45 years since *The Phenomenon of Man* was written, and as we look back, we have a new realization of the profound effect of humans on the planet. Extinct and endangered species, global atmospheric contamination, a threatened ozone layer, and a spreading area of urban development all testify to the reality of the profound changes that reflective thought has brought to the world. We are perhaps more sympathetic to viewing the origin of the noosphere as a cataclysm that has changed the rules of evolution, removing them from a purely biological context and forcing new methods of analysis. In addition, the very way in which we view physics has changed considerably in the last 50 years, making the human mind much more a part of the theoretical structure of that subject. As physicist Eugene Wigner has written about quantum mechanics, "It will remain remarkable, in whatever way our future concepts may develop, that the very study of the external world led to the conclusion that the content of consciousness is the ultimate reality." The physics that underlies our biology, in a reductionist sense, has come in recent years to see its own foundation in reflective thought, bringing Teilhard's views of the transition to the noosphere closer to normative science.

After the discussion of the transitions, *The Phenomenon of Man* drifts into realms of theology and mysticism, forcing the scientist, as a scientist, to part company with Teilhard's thoughts as being totally beyond the discipline of biology. Nevertheless, within his notion of major transitions there appear to be major insights. The last of these radical changes, the rise of reflective thought, is the least understood. It is a concept that we may well probe further, for it promises to tell us much about our humanness.

Mayonnaise and the Origin of Life

The volume in my kitchen on the art of French cooking has a basic mayonnaise recipe that lists only three constituents: vegetable oil, egg yolk, and vinegar. A recipe for mixing up a living cell would be vastly more complicated, but the microstructures of both cell and salad dressing depend on a class of molecules of the greatest importance in all living processes. These ubiquitous chemicals are designated the amphiphiles. Since that noun occurs neither in the *Oxford English Dictionary* nor in the *Webster's New Collegiate Dictionary*, I feel safe in assuming it is not a household word. Actually, it was necessary to go to *Chemical Abstracts* to confirm the usage. What is surprising about the relative obscurity of the amphiphiles is the fact that as requisites for life they are just as important as proteins or DNA.

The word itself is a compound of *amphi*, meaning "both kinds," and *philos*, "to love." The two kinds referred to here are oil and water, for these molecular structures have one end that is attracted to oil and another end that is attracted to water. The presence of amphiphiles in the egg yolk is what makes the smooth, homogeneous mayonnaise sitting in my refrigerator belie the old adage that oil and water don't mix. The egg material contains lecithin, a phospholipid molecule with a water-soluble phosphate moiety and a lipid portion that dissolves in oil. Lecithin therefore occupies the interface between the oil droplets and the surrounding vinegar and thereby produces a smooth emulsion—pleasant, tasty, and fattening.

This discovery of the use of egg yolk to blend oil and water goes back to some time before 1807, when the word *mayonnaise* first appeared in print in *Le Cuisinier Impérial* of Viard. The origin is obscure, but there are those who trace it to a feast celebrating the capture of Port-Mahon in 1756 (*mahonnaise*). Others credit the chef of the epicure the Duke of Mayenne (*mayennaise*). Some philologists favor the verb *manier*, "to stir" (*manionnaise*), while another school leans toward *moyeu*, an old French noun meaning "yolk" (*moyeunaise*). Leaving that scholarly matter unsettled, we do note that salad dressings were by no means the first technological use of amphiphilic emulsifiers. We can trace the literature all the way back to Pliny the Elder, who wrote about boiling goat's tallow and causticized wood ashes to make soap. Pliny was unaware that soap solubilizes oily dirt by surrounding droplets with bifunctional molecules that allow them to be freely suspended in water.

The two active ends of an amphiphile are designated hydrophilic (water loving) and lipophilic (oil loving). The basic physics of these phenomena is well understood. All matter is made up of heavy, rather stationary nuclei and much lighter, highly mobile electrons. Lipophilic molecules have a high degree of symmetry of the electrons about the nuclei, both in the presence and in the absence of an electric field. They therefore have a low dielectric constant, in contrast to hydrophilic molecules, which are electrically asymmetric. These properties lie very deep within the foundations of molecular physics.

Amphiphilic molecules are a fundamental structural component of all living systems and are found most commonly in the membranes of cells and organelles. As our knowledge of biological membranes has increased, we have become aware that they are almost always built out of amphiphiles in a planar array two molecules thick. These bilayers have the oily parts of the molecules in the interior, while the "water loving" portions are on the surface to maximize their contact with aqueous surroundings.

Unlike the mayonnaise particles, which are spheres of oil held in a watery environment by amphiphiles, membranes are closed shells made entirely of amphiphiles. The oily part of the bilayer separates the aqueous interior from the exterior of the structure. Plasma membranes, chloroplasts, mitochondria, and endoplasmic reticulum are all made in the manner described above. We could not envision a cell without these split-personality molecules, which, being unable to decide between oil and water, are consigned to exist permanently at interfaces.

The emerging preeminence of amphiphiles in cellular structure has led a number of students of the origin of life to take a hard look at the possible role of these molecules in biogenesis. Some contemporary speculation suggests that the first step toward cellular life consisted of the formation of vesicles with bilayer membranes. In 1971 A. C. Lasaga and co-workers suggested that before the advent of life the early oceans were covered with a layer of hydrocarbons (*Science* 175:53). This is going to come as a great shock to environmentalists, but that proposal, based on some experimental data, postulates a primordial oil slick that "divided the waters which were under the firmament from the waters which were above the firmament." Other experiments have shown that the ultraviolet irradiation of such hydrocarbons floating on water can lead to the photochemical production of amphiphilic molecules. At a very early stage in the history of the planet conditions may have existed for the synthesis of membrane precursors. From that point it is not difficult to envision the formation of small membrane-bounded bags of organic molecules, the forerunners of later cells.

From a more humanistic point of view, individuality entered the world when the first membrane fragment wrapped itself into a closed shell and separated the interior components from the rest of the universe. A group of chemicals in solution, regardless of how complex, cannot achieve individuality; components are always being dissipated by diffusion. Therefore a sine qua non for the existence of individuality is a diffusion barrier, an envelope to partition a part of a system from its surroundings. This requires a phase separation, and the hydrophilic-lipophilic distinction seems like the clearest way to make such division in a world that is dominated by water.

The existence of amphiphiles stems from the very basics of electromagnetic theory. At a higher level these molecules reach up toward the beginnings of biological individuality. They are of vital importance now and appear to have been equally important in the salad days of our planet. We are clearly dealing with most significant pieces of biological apparatus. If you have not been informed of them before, ignore that journalistic omission; you will hear much of amphiphiles in years to come.

Beetles, Ecologists, and Flies

In an anecdote popular among evolutionary biologists, J. B. S. Haldane is said to have met once with a group of British theologians. They asked him what he was able to learn about the Creator from a study of his creation. "An inordinate fondness for beetles," the noted biologist reportedly replied. That story may come as no surprise to gardeners who encounter these insects in great abundance, but its real meaning lies not in the number of beetles, which is vast, but in the number of species of the Coleoptera, which is larger than that of any other animal order. Next to the beetles in species richness come the flies, or Diptera, and together these two groups have more recognized species than the entire remainder of the animal kingdom. This somewhat surprising information leads to the inquiry of why there should be so many species of small insects. Scientists seek an answer more empirically rooted than Haldane's whimsical reply to his ministerial associates.

The richness of animal variety as a function of size was the main theme of a 1977 report by ecologist Robert May to the Royal Entomological Society. Under the heading "Dynamics and Diversity of Insect Fauna," May reviewed a series of studies on the number of species of all land animals in different size ranges. The results that he presented can be summed up as follows:

Length of animal	Number of species
Less than 0.01 inches	20,000
0.01 to 0.1 inches	220,000
0.1 to 1.0 inches	600,000
1.0 to 10.0 inches	20,000
10.0 to 100.0 inches	1,500
More than 100 inches	10

We find no animals tinier than the smallest protozoans at 0.001 inch or larger than the African elephant at 150 inches. The vast majority of animal species are around half an inch or less, which is the middle range of insect lengths. The question of why there are so many kinds of flies and beetles becomes an inquiry as to why there are so many varieties of animals in this size range.

Discussions about the size of animals have in recent years centered on the concept of a niche, an idea widely used by ecologists in explaining interactions of organisms that occur in the field. The description of a species niche consists of two parts: where organisms live and how they make a living. Consider a few examples: grass eaters on semiarid plains, fruit eaters in tropical forests, plankton eaters in Arctic seas. A niche is specified by the environmental conditions within which a species gets its food, evades predators, and carries out physiological and reproductive functions. The description includes both physical parameters of the habitat, such as temperature and humidity, and biological parameters defined by other species living in the same surroundings. The important biological features center on who eats whom, as well as the other ways in which organisms can affect each other.

Modern ecologists have two general rules governing the number of species in a given habitat. The first, called the principle of competitive exclusion, asserts that only one species can occupy a given niche. If there are two species living in exactly the same place and eating exactly the same food, one will outcompete the other. The loser in the struggle for survival will be driven out or become extinct. The number of species, then, is limited by the number of available niches. A second generalization states that nature tends to fill an empty niche by evolutionary change of an existing species or the immigration of a species from elsewhere. In the famous case of Darwin's finches, niches on various islands of the Galápagos were filled both by the arrival of birds from neighboring islands and the evolutionary change of beak structure and behavior once the finches had arrived.

In 1958 ecologist G. Evelyn Hutchinson wrote an essay entitled, "Why Are There So Many Kinds of Animals?" He concluded that the richness of plant life created a vast number of possible niches, leading to the proliferation of animal species. The plants, in their own struggle to survive, had diversified by evolving structures that provide many microenvironments. We must now inquire why these microenvironments favor animals of a certain small size. The principle of exclusion would argue that there must be more different kinds of niches for these middle-sized animals than for either larger or smaller fauna.

First consider the case of very large animals. African elephants have annual migrations seeking food and water that cover hundreds of miles in distance and an area of thousands of square miles. Whales swim thousands of miles through the world's oceans to provide for their needs. Such large mammals have niches that occupy appreciable portions of the earth's surface and available food supply. Hence there can be very few such niches because of the limited size of the planet's surface, and accordingly, only a small number of large animal species are found. In the intermediate-size range there can be a large number of microenvironments because the plant structures—leaves, branches, roots, and so forth—are the same order of size as the animals, creating a structural richness and giving rise to many niches. The shady underside of a leaf is a different environment from the sunny upper side. Eaters of thick, older leaves at the bottom of a plant have a diet different from animals that feed on the tender upper leaves. Intersections of different branches produce varied web-building locales. If one focuses on zones from a hundredth of an inch to several inches, a vast number of such descriptions is possible, hence a richness of niches.

But as we focus on even smaller animals, a new constraint enters the picture. Below a certain size, it becomes very difficult for organisms to survive in air because of dehydration. The loss of water by evaporation depends on surface area. As objects get smaller, there is more surface per unit volume. Thus, for example, a spray of water evaporates far more quickly than an open bucket containing the same amount of liquid. As a result, some very small land animals, such as tardigrades, tend to favor such moist niches as wet moss. Even so, these creatures have had to evolve special mechanisms to survive dehydration when the humidity drops and the leaf surface dries. Tardigrades can exist in a dried, inactive state and then revive when they once again become wet.

Because of the drying problem, most very small animals tend to be aquatic. Their niches are characterized by such chemical features as acidity and salt concentration rather than the structural features provided by land plants. Since these niches often extend over considerable areas, any one species will tend to be quite populous. But because there are many fewer chemical parameters than there are possible structural ones, there are many fewer distinguishable niches for water dwellers. This results in a relatively limited number of species at the smallest end of the range.

The maximum number of niches available, then, is for animals whose size is somewhere between tiny water dwellers, such as protozoans and rotifers, and the more sizable land-dwelling mammals, birds, and reptiles. It is a size range most suitable to the physiology and anatomical features of insects. For ecological reasons nature seems to have an "inordinate fondness" for animals that can make a living crawling and flying around the small spaces in trees, shrubs, and other vegetation. Beetles and flies obviously excel at that kind of activity, and they have consequently emerged with a greater species richness than all other forms.

A Leap of the Imagination

One of the most dramatic and startling legends from classical Mediterranean civilization informs us that the philosopher Empedocles jumped into the fiery crater of Mount Etna in order to offer proof of his divinity. The experiment, alas, failed, and history stands mute as to the response of those devotees who climbed to the peak of the mountain with their teacher.

What have endured from the life and times of this extraordinary thinker are some 400 lines of the poem *Peri physeōs* ("On Nature"). In the beginning of Western intellectual life no sharp distinction existed between arts and sciences; a poem and a scientific treatise were often one and the same work. Within this framework we can comprehend that Empedocles was hailed as philosopher, statesman, poet, religious teacher, and physiologist. Nowadays he is best remembered for his major theoretical insight, a leap of the imagination in which he postulated that all matter was made up of four basic and essential elements: earth, air, fire, and water. All processes were then to be understood in terms of the mixing, unmixing, and remixing of the fundamental stuff of the universe. These reactions were governed by the forces of attraction (love) and repulsion (strife).

The simplicity and deep explanatory power of Empedocles' theory were persuasive to the thinkers of the Athenian academies. Both Plato and Aristotle, as well as their numerous followers and successors, adopted the four-substance point of view in one or another of its forms. The chemical paradigm of the Golden Age of Greece was the legacy of this charismatic Sicilian scholar, whose career seems to have been prematurely terminated by a religious theory less successful than his scientific thought.

With the decline of classical civilization the works of Empedocles, along with most of the learning and lore of Greece and Rome, passed into relative obscurity, becoming the province of a few scattered scholars. Some 2,400 years after the death of the founder of the earth, air, fire, and water doctrine, a new science was founded that curiously echoes this early division of matter into its components. The beginnings of modern geology in the eighteenth and nineteenth centuries led to the introduction of the concept of geochemistry by a Swiss chemist in 1838. When the first modern book on this subject appeared in 1908, it was clear that the science not only had to include the rocks and mineral materials but had to consider the interactions and properties of the planetary air and water. Thus the geologically studied portions of the planet were classified as lithosphere (rocks and crust), hydrosphere (oceans, rivers, and lakes), and atmosphere.

Although these three geospheres contained over 99.9% of the surface material, they proved inadequate to describe the chemical changes constantly occurring on the earth. To focus on many processes of growth and decay, it was necessary to consider another catalytic component, the living organisms, whose extensive chemical activities were appreciably out of balance with their relatively small mass. This fourth domain was finally given a proper name in 1929 when Russian geochemist V. I. Vernadsky published his book *La Biosphère*. The tetradic organization of matter envisioned by Empedocles was restored to its position of prominence in an altered but still recognizable form.

By 1950 the subject matter of the new science had solidified and was recorded in *Geochemistry*, the now classic treatise by Finnish scientists K. Rankama and T. G. Sahama. Prominent in this work was the division of surface material into four spheres—lithosphere, atmosphere, hydrosphere, biosphere—and the characterization of each of these major constituents.

The lithosphere is composed of the solid portion of the surface of the planet; the stones and clays and soils and bedrock upon which they sit are parts of this whole. It is the largest of the four spheres and clearly corresponds to the "earth" of Empedocles. Specific reference to the great land masses are to be found in the lines of *Peri physeōs*.

The atmosphere of contemporary geochemistry is almost certainly identical with the "air" of the classical scholars. We now have a much greater knowledge of its upper levels, including the stratosphere. Airplanes and orbiting satellites have provided a wealth of experimental information that Empedocles could only have gotten by interviewing the legendary Daedalus. The knowledge of the stratosphere available from Icarus's observations was sadly lost when his wax wings melted and he fell into the sea and drowned.

The hydrosphere of modern science covers 70.8% of the globe and includes the oceans, rivers, lakes, and glaciers. Although dominated by water, it is not quite identical to the basic substance of Greek philosophy. The classical "water" is conceptually closer to the compound hydrogen oxide (H_2O), which is a major part of but not the entire hydrosphere. In addition, water molecules exist in the atmosphere as vapor, are a dominant component in organisms, and contribute to clays and hydrated minerals in the lithosphere. Thus, we have "water," both as the chemical compound in all four geochemical spheres and as the closely related hydrosphere that contains most of the planet's H_2O.

Lastly, we find the biosphere that we identify with the fire of the classical tetrad. Although this analogy requires some imagination, it should be remembered that almost all combustion known to the Greeks involved biological products, such as wood and fat, chemically combining with oxygen, itself produced by the biological process of photosynthesis. The second kind of fire they experienced was the sun, which provided the radiant energy for plants to store as chemical compounds. The third kind of fire was volcanic in origin, and this only became biological when Empedocles chose to merge his person with the molten lava. If we are stretching a point in associating biosphere with fire, both are clearly major elements in the chemical processing of solar energy, and that is the main issue from the earth scientist's point of view.

In any case, the modern theory of geochemical spheres allows us to describe the demise of the great Empedocles in a way that is congenial to his view of nature and imparts a bit of immortality, if not divinity, to his remains. For upon contact of his body with the molten basalt, we may assume that about 50 kilograms of water was released. This is now mixed with oceans, clouds, and rivers, so that the beaker of hydrosphere now before me contains about 100,000 atoms of the early savant's oxygen along with 200,000 atoms of his hydrogen. The philosopher's carbon and nitrogen quickly joined the atmosphere as carbon dioxide, methane, ammonia, and molecular nitrogen. These atoms have since participated in the great global ecological carbon and nitrogen cycles. The nonvolatile constituents remained in the lava, which eventually rolled down the slopes of Mount Etna, forming the hardened basalt, a part of the lithosphere. These minerals still leach out to nurture the growing crops of sunny Sicily. The atoms of Empedocles persist, as does his theory that the matter of the earth has a fourfold division into its component parts.

Two Views of Life

Scientists working at the laboratory bench or in the field spend little time pondering philosophical issues. Nevertheless most biologists maintain in the backs of their minds a concept of "life" that motivates their work. We have recently come to view life as a property of very special molecular arrangements called cells. Within this framework the essence of aliveness can be studied in the laboratory, analyzed, and reduced to its atomic fundamentals. This view led to molecular biology, including a precise knowledge of the roles of DNA, RNA, and proteins in cell growth and replication. It underlies genetic engineering and other attempts to apply biology to practical problems.

There is a different and equally insightful answer to the question "What is life?" It is now emerging among scientists interested in the geochemical history of the earth. For this group, life is a property of a geologically active planet whose elements keep recycling under the driving forces of solar radiation and mechanical energy from processes within the mantle and core. This viewpoint has emerged from the cooperation of ecologist, geologists, meteorologists, and oceanographers, and consequently its formation has taken a long time. Findings in all these fields confirm that the life of any single organism is part of a larger-scale process involving the "metabolism" of the whole earth. Continued biological activity is then a planetary property, an interrelationship of organisms, atmosphere, oceans, and continents, all of which are in some sense alive. Without the part played by each of these components, life as we know it would, on a geological time scale, grind to a halt.

This global view stresses the fact that earthly life began billions of years ago; grasping its essence must include an understanding of how it has persisted for so long. A second incentive for the planetary approach follows from our inability to separate geochemical and ecological cycles, because they are so deeply intertwined. These major processes on the earth's surface are powered by a constant flow of photons leaving the sun and traveling with the speed of light to our planet. This energy drives the chemical reactions of the atmosphere, all of photosynthesis, and most meteorological processes.

Beginning in the upper stratosphere, photons in the ultraviolet portion of solar radiation excite a series of chemical reactions that produce ozone and various compounds of nitrogen from normally occurring atmospheric molecules. The ozone shields the lower atmosphere from too much ultraviolet radiation, and the oxides of nitrogen fall to the surface in rain and enter the biological nitrogen cycle. Thus the photochemistry of the outermost gas layer is linked to the biology of the planet. Since oxygen in the atmosphere comes largely from living organisms, the linkage is reciprocal. This much of the new view has been a central theme of James Lovelock's Gaia hypothesis, which deals with the self-regulating properties of the planet earth.

On the land and in the waters of the earth a small fraction of sunlight is utilized by green plants and pigmented bacteria to carry out photosynthesis, which links carbon dioxide from the atmosphere with water from the surface to make sugar and oxygen. The energy locked up in these compounds is later released by metabolism, and it powers most biological activity. In these chemical reactions the energy appears as heat and is eventually lost in the form of infrared energy radiated back into space.

During plant and bacterial growth, phosphorus, sulfur, calcium, iron, and other minerals are removed from the soil, and nitrogen fixation takes place. The great nutrient cycles consist of the incorporation of chemical elements in the process of growth, followed by their release during respiration, excretion, and decay. These processes are carried out by plants, the herbivorous animals that eat them, the carnivores, and lastly those countless tiny organisms of rot and decomposition. The physical and chemical properties of soils and waters greatly influence the creatures that grow within and upon them. In return the biota depend on the features of their surrounding habitats.

The major portion of solar energy reaching the earth warms the surface, evaporates water, and otherwise drives those phenomena we recognize as weather. Associated with these meteorological activities are the erosion of rocks and leaching of soil by wind and waves and water. This constant weathering keeps carrying sediments and dissolved minerals to the rivers and then into the oceans. If this process, acting alone, had continued for eons of time, the land would have been impoverished in vital nutrients and made unfit for habitation. Since life has persisted, some method of land regeneration must be at work. In addition, if weathering were not countered by continued mountain-building and geological uplift, the entire planet would have become a uniform mud flat devoid of terrestrial features.

In the depths of the seas a steady rain of tiny rocks, insoluble minerals, and nondegradable residues of dead organisms falls down to form the deep oceanic sediment. This layer inexorably grows, and in the absence of geological processes, such a large fraction of vital chemical elements would pile up on the sea floor that all life would end because of a shortage of essential nutrients. The persistence of life on earth presented a great paradox to geologists because sediment formation should long ago have stripped all habitable areas of life-sustaining elements.

During recent years the geologist's study of plate tectonics, which deals with such things as the motion of the continents, has resolved the most difficult problems of continuity of life. It has been discovered that when continental plates slide over the ocean floor, bottom materials, including sediment, are driven down below the crust into the magma, a layer of partially molten rock. Earth scientists call this process subduction. Within the magma, great physical and chemical changes take place, altering the materials that eventually rise to the surface again in mountain formation, volcanic activity, and other geological processes. The chemical elements that have been lost as sediment return as gases, rocks, and lava. The rocks and lava weather to become soil, making their matter available once more to living organisms. The very crust of the earth has a cycle all its own, continuously replenishing the planet for plants and animals. The time it takes to complete this cycle can be guessed at from the fact that no part of the ocean floor is older than 200 million years.

In addition to these subtle but massive geological activities there exist in the oceans of the world deep-sea currents sweeping up dissolved nitrogen from the sediments and bringing it back to the surface in cold, nutrient-rich water. Such an upwelling occurs in the Pacific off the coasts of Peru and Ecuador and is responsible for that region's rich fisheries. The circulation of ocean currents is another participant in the global living process.

In 1913 Harvard physiologist Lawrence Henderson wrote a widely read and frequently quoted book, *The Fitness of the Environment*. Speaking as a scientist, he argued for the overview that the physical and chemical conditions of the earth are uniquely ideal for life. The outlook that is emerging today goes much farther and maintains that the environment and living organisms are bound, inseparable parts of one set of linked planetary processes. Within this context the sustained activity of the global "biogeochemical" system is more characteristic of life than are the individual species that arise, flourish for a time, and disappear in the course of evolution.

How are the global and laboratory-centered views of life to be combined in a unified theory? How are the specific molecular mechanisms of DNA, RNA, and protein related to the massive cycles of carbon, nitrogen, sulfur, and phosphorus? The answers remain elusive, but now more than ever we can see that the questions must be addressed. When and if we establish an understanding of the relations between these complementary views of life, we will have a much better idea of who and what we are. In the meantime the global view might induce us to love our planet a little more—volcanoes, earthquakes, and all.

Bird What?

Ogden Nash, the master of the strange rhyme, once penned the immortal lines:

There is nothing in any religion
That forces us to love a pigeon.

Good poetry may on occasion be bad theology, but Nash's lines do serve to remind us of the countless statues, park benches, straw hats, white suits, and print dresses that have been unesthetically besmirched by the physiological products of incontinent birds. However, life is often not what it seems, and we move from bird excrement to one of the wealthiest nations on the surface of the earth.

When queried as to the country with the largest per capita income, most people think of Sweden and the various oil emirates. Few would remember to include in that select list Nauru, a tiny island republic lying in the Pacific Ocean at longitude 166° east, just below the equator, and due north of New Zealand. At the center of the island are mounds of high-grade trisodium phosphate that bring a high price on the world fertilizer market. These substantial quantities of mineral are the sole source of the nation's income. Of course, compared with the industrial giants, the cash flow of Nauru is small. But since it is divided among only 4,300 citizens, simple arithmetic reveals the reason for the high per capita figure.

The extensive use of phosphate fertilizer points up the importance of phosphorus in the nutrient balance of living organisms. In a world dominated by a strongly oxidizing atmosphere, phosphorus almost invariably exists as phosphate (PO_4^{3-}), and that chemical group persists in biochemical molecules. Membranes are made of phospholipids, nucleic acids are held together by sugar phosphate bonds, and at the core of energy processing lies the essential compound adenosine triphosphate. Because of washout from soil and precipitation in waters, phosphorus is often the limiting element in natural systems. This is dramatically demonstrated by the great surge of growth that accompanies the addition of phosphate to certain lakes. The resulting eutrophication (nourishment) is hardly regarded as an ecological plus by fishermen, swimmers, and lakeside residents.

Where then does the great natural bounty of Nauru come from? The process begins at the icy surface of the Norwegian Sea, where water is cooled and convectively carried downward, starting the deep oceanic circulation. In traversing the benthos over many years, these waters become enriched with nitrogen and phosphorus that have precipitated into the depths. At certain locations in the Pacific and Indian oceans these currents upwell, providing rich nutrients for myriad algae to be fruitful and multiply. These tiny primary producers are food for communities of zooplankton, which are dominated by the ever-present copepods. The grazers are in turn eaten by larger organisms, including fish. An ecological food chain is established, transferring phosphorus and other minerals from the ocean bottom into teeming surface life. At the top of the food chain, both physically and hierarchically, are the birds that swoop down into the ocean and snatch the fish for food. The great cycle of life carries phosphate in a biologically fixed form to countless pelicans, boobies, gannets, cormorants, and frigate birds.

Part of the mineral intake of the predators is utilized for growth, while the remainder is transformed into guano, nutrient-rich excrement. In those areas of the world where birds pause to relieve themselves, the ground becomes covered with organic material. If climatic conditions are right, the residue accumulates into extensive deposits. The good fortune of the Nauruans is that many, many generations of migrating birds have chosen their island to rest and nest.

The economic importance of guano cannot be overestimated. During the mid-nineteenth century 180,000 tons of this fertilizer were harvested each year from the western coast of Peru, a locale that borders a particularly rich oceanic upwelling area. The bird excrement has even found its way into the history books via an unusual piece of legislation. In 1856 the United States Congress passed the Guano Act, providing the legal basis for the annexation of the Pacific atolls of Baker Island, Howland Island, and Jarvis Island.

When the bird feces age under the right meteorologic conditions, bacterial action and weathering result in the loss of gas-phase materials, such as carbon dioxide, nitrogen, ammonia, and hydrogen sulfide. There are no biochemical pathways for converting phosphates and cations to gas-phase molecules; hence the fecal matter is slowly transformed into the residual trisodium phosphate that now enriches the fortunate citizens of Nauru.

The residents of this lovely Pacific island lived in ignorance of their great wealth when the island was visited by British navigator John Hunter in 1798. After many visits by whalers and other seamen the territory was annexed by Germany in 1888. Exploitation of the phosphate reserves began in 1906. Following the vicissitudes of international events Nauru became British, Japanese, a United Nations Trusteeship, and finally, in 1968, an independent state. Today Nauru is an island republic whose citizens all share in the bounty deposited by the avian visitors. Lest this statement start a mass migration, we note that the Nauru constitution defines citizens as members of the Nauru community on January 30, 1968, children of Nauruan citizens or of Nauruan citizens and Pacific Islanders, wives or former wives of Nauruan citizens, and other granted citizenship by acts of Nauru's parliament.

At present over a million tons of phosphate are mined each year, and the supply is expected to run out in 1990. This is truly a case of endangered feces, and the Nauruans are planning for the resource-depleted future by investing their money in an airline, a shipping company, and other commercial enterprises. The republic owns a 32-story skyscraper in Melbourne, Australia, called Birdshit Towers by the local wags. The holdings are kept in a constitutionally established Long Term Investment Fund, from which monies cannot be withdrawn before the depletion of the island's phosphate resources. The Nauruans may cheerfully look forward to never ever ever having to work for a living. For those who believe that success consists of being at the right place at the right time, we can only suggest buying an island, mounting a large bird feeder, and waiting patiently. For those whose puritan work ethic is offended by this story, I can only reply that I am simply reporting facts; moral judgments will be left to others.

The preamble to the Nauruan constitution begins with the words: "Whereas we the people of Nauru acknowledge God as the almighty and everlasting Lord and the giver of all good things. . . ." Our story began with Ogden Nash's idiosyncratic views of God and bird, and it ends with the proclaimed faith of the Nauruans. Whatever one's religious views, it must be agreed that the blessings of Nauru have come from on high.

SECTION TWO

THE ROOTS OF WISDOM

Thoughts about Medicine and Dentistry

The Roots of Wisdom

This [1982] being the one-hundredth anniversary of the birth of President Roosevelt, we old-time residents of Dutchess County, New York, like to sit around our cracker-barrels and swap our Franklin Delano stories. Most of them are pretty well worn by now, and it was a special treat to hear a genuinely new tale from my favorite dentist, now living in retirement in Florida.

In time and place we go back to June 1932 in Poughkeepsie. The phone in the dentist's office rang, and the voice on the other end was that of John E. Mack Sr., requesting an immediate appointment. Judge Mack was not a man to be treated lightly. One of the county's most respected jurists, he was also a politician of some weight, a former district attorney, and a friend and confidant of then-Governor Roosevelt.

Mack appeared shortly and told his sad story. He was suffering from failing eyesight. His doctor had been unable to make a diagnosis and, suspecting an infection, had recommended a dental consultation. Although the Judge was not a patient of my dentist, several members of his family were, and no questions were asked about why he was not taking the matter to his own practitioner. The Judge was a formidable presence, and one did not, as a rule, question him. A full-mouth series of x-ray films were shot, and an appointment was made for the following morning.

The pictures revealed a severe deep abscess, due to the incomplete removal of the roots in an earlier extraction of a third molar. The patient was informed that there was no assurance that treating the abscess would cure the eye problem, but good health mandated action in any case. The Judge whipped out another set of X rays and asked, "Could you reconcile your diagnosis with these X rays from the hospital with the accompanying diagnosis of no dental infection?" A comparison of the pictures showed that the hospital films of the molars had not gone far enough down to detect the decaying root. The Judge looked at the films and made his decision. "When can you begin treatment?" "Right now," was Mack's response. "I'm going to have to call Franklin and tell him I can't speak for him," muttered the Judge, half to himself. In the summer of 1932 in Poughkeepsie there was no doubt who Franklin was.

The speech being referred to was no ordinary political oration, but the nomination of the Democratic candidate for the presidency of the United States. John Mack was scheduled to leave for Chicago shortly for the most important moment in his political life. His illness threatened to deprive him of that triumph.

The dental work, which began immediately, was a long and rather difficult procedure, getting down into the bone and cleaning out the abscess. In the days before antibiotics, success depended on removing all traces of necrotic tissue, and the dentist did his job well. Once the sutures were in, the Judge left, again saying, "I've got to call Franklin."

The following day Mack stopped by to have his jaw examined; he reported feeling a little better. Twenty-four hours later he noted that his eyes had begun to improve. Several days afterward, on June 30, he was ready to place before the convention the name of Franklin Roosevelt. Historian Arthur Schlesinger Jr. referred to Mack's address as "uninspired," but neither Schlesinger nor anyone who was in Chicago that day was aware of what had taken place shortly before in a dental office in Poughkeepsie. In his 12 years of practice the dentist had never seen such a dramatic cure of an infection-associated disease following the treatment of an abscess. The Judge's eyesight ultimately returned to normal, and he continued his brilliant legal career. He, of course, became a lifelong patient of my dentist.

When the Judge was queried about why he had sought a second opinion after the first set of films, he replied, "A man would be a boob not to follow his physician's instructions, and he told me to go to the hospital and get a full set of X-rays. But nobody told me I couldn't also go to somebody else." That was really a nonanswer, but it's all we have to go on.

I listened to this story with fascination, for early in November of 1932 I had stood in a crowd before the balcony of the Nelson House in Poughkeepsie, thrilled to see the man I somehow knew was about to become president of the United States. My father had taken me by the hand to witness history in the making. We had walked down Cannon Street to the corner of Market. The crowd was so dense that we could go no further. I was too short to see anything over the shoulders of those jammed together in front of me, so Dad lifted me up to the base of the Grecian column of the Farmers and Manufacturers National Bank building, directly across the street from the hotel. On the platform with the candidate was John E. Mack. The scene is very clear in my mind, and I can recall the closing words of Roosevelt's speech: "I am not asking you to vote for me, but I am asking you to vote for the man who the majority of Americans will vote for." I also remember that Roosevelt, supported by two of his sons, moved slowly to the podium.

John Mack's presence that night and earlier in Chicago did not play a decisive role in history, but the story reminds us of the large part that the health or sickness of leaders can have in influencing the course of events. The health of United States presidents has been reexamined in some detail by Dr. John B. Moses and Wilbur Cross in *Presidential Courage* (W. W. Norton & Co., 1980). Woodrow Wilson's stroke is probably the most significant presidential disability from a national point of view.

Wilson was ill on and off for his entire adult life. A series of medical problems plagued him during his term of office, culminating with a paralytic stroke in October 1919. His wife and kitchen cabinet ran the office and engaged in a conspiracy to keep secret Wilson's disability. American support for the League of Nations faltered, with profound effects on the entire world.

Franklin Roosevelt's health during his fourth term has been a matter of considerable speculation among historians. At the famous Yalta Conference the American president met with Churchill and Stalin to set the map of postwar Europe. A sick, tired FDR had difficulty focusing on the issues. Many feel he gave away too much. Winston Churchill's physician estimated that Roosevelt had two months to live. Sixty days later a massive cerebral hemorrhage ended the life of the man whom John Mack had nominated in Chicago 13 years earlier.

It was a long way from the balcony of the Nelson House to Yalta. Roosevelt's career, begun among the farmers of Dutchess County, ended in sickness, negotiating the boundaries of modern Europe. We perhaps tend to underestimate the historical role played by the good health or illness of leaders. John Mack's abscessed jaw serves as a reminder of the importance of both the pathological condition and the care given.

My Mouth Shall Praise

At the risk of being regarded as mildly idiosyncratic, I would like to explain why time spent in the dentist's chair during the past 30 years has, all things considered, been enjoyable. It is not masochism or insensitivity to pain or some perverse oral gratification that comes from a mouthful of whirling and grinding instruments; rather it is the intellectual and esthetic pleasure I get from observing an accomplished and dedicated expert at work. My dentist fashions a preparation with all the care of Benvenuto Cellini shaping a filigree of gold or of Michelangelo bringing forth a human shape out of a block of marble. Each step is done with great deliberation and attention to a final product that reaches for perfection. This concern with quality is so gratifying to watch that it full well overcomes the admitted unpleasantness of having one's teeth and jaws the objects of mechanical manipulation.

Thus, I repose in tranquility as the residuum of my upper left first molar is being shaped for a cap. It seems a good time to think about the subject of quality, for equipment in the mouth makes speech inelegant, if not downright sloppy. There has been a lot of talk around lately about where quality in work performance has gone in the modern world. The last few months have brought personal examples: a surly waitress, a pediatrician too busy to talk to a mother about a problem child, an air freight company employee who couldn't care less about an overdue package of frozen research materials, and a car salesman unable to answer the simplest questions about his wares. Each reader can doubtlessly supply a similar list, and the aggregate represents for all of us a way of life that is less than it might be.

The sociologists have much to study about the reasons for this loss; industrialization and population pressure are clearly contributing factors. Being unequipped to undertake a formal social science analysis, I must get my teeth into the subject in a different way. The first step in reversing the decline in quality of life is to praise that glorious attribute whenever we encounter it. To make quality a goal, we must first firmly incorporate it into our ideology, and that is surely the work of poets, writers, and philosophers, as well as others among us.

All of which brings me back to my dentist, who can hardly be ignored, since he is now pressing down on a hardening matrix to take an impression of the precisely shaped remains of my tooth. You see, there is a special reason for my thinking about his professional services today. After 55 years of practice he is about to retire. Although I have been a patient of his for only 30 of those years, the thought of looking for another dentist is causing me great consternation. Having been so thoroughly spoiled, I will have great difficulty finding anyone to meet my standards. Like Diogenes looking for an honest man, I envision myself, lantern in hand, wandering from office to office, looking for a quality dentist. In the end, it may be necessary to give up having dental problems.

While the impression sets, the concept of "55 years" begins to solidify. Although his grandchildren affectionately tease our dentist about taking early retirement, 55 years is a long time. No doubt many professionals emerge from school as young idealists with a commitment to quality. It is quite another matter to maintain that dedication through more than half a century of active practice, with all its difficulties and frustrations and the allures of taking the easy way out. Twenty thousand days of pursuing a goal requires a strong guiding light.

In the dentist's chair one is deprived of a dictionary, so the problem of defining *quality* takes on a transcendent aspect. The devotees of excellence I have known have rarely been the wealthiest practitioners, although they have usually succeeded. The kind of performance we are discussing takes great care and much time. As a result, such perfectionists will have a lower gross productivity than others. But that may be part of the essence of what we are looking for, a willingness to trade tangible quantity for the somewhat more ephemeral and elusive goal. A rare genius like Shakespeare may be able to achieve both ends, but for most mortals quality must be purchased at the price of total output.

I am now being treated to a splendid example of the above. My dentist is engaged in fashioning a tooth that will be worn for only two weeks, while the laboratory processes the permanent cap and its coating. Yet he is so delicately shaping the temporary prosthesis that it is emerging as an object of beauty itself. It will shortly be thrown away, yet this professional cannot, as a matter of character, allow it to be less than the best of which he is capable. (When I subsequently get to the dictionary, I find that the definition of *quality* I am discussing is given as "excellence." The example better expresses the connotation of the word.)

We cannot dwell on the past. In spite of future personal inconveniences, it is clear that my dentist's retirement is richly deserved. He will go on seeking quality in the many aspects of daily living. I hope he will somehow have the opportunity to convey his attitudes to a younger generation of dentists so that they will be encouraged to seek in their own work the kind of approach that has sustained him those 55 years. For my part, my retirement present to him is a promise to praise quality whenever I see it. Those of us with literary soapboxes not only must be concerned with the intrinsic quality of our own work but must be ideologically committed to excellence in all fields of human endeavor.

The temporary cap is now in place, and I leave the office with a strange feeling in a novocaine-numbed jaw and a peculiar lightheadedness engendered by such a concatenation of philosophical and pragmatic thoughts about the practice of dentistry. At least it is now clearly understood why time spent in the dental chair for the last 30 years has resulted in a paean rather than a pain.

Between Gargoylism and Gas Gangrene

And scorne not Garlicke, like some that thinke
It only makes men winke, drinke, and stinke.
—Sir John Harington, *Englishman's Doctor* (1607)

Every so often a writer gets the scent of a really spicy story and is impelled to follow his nose and dig up the facts. I thus found myself in the library once again, poring over the pages of *Index Medicus*. And true to form, that repository of knowledge did not disappoint me: cataloged between GARGOYLISM and GAS GANGRENE was found the object of my research, the subject entry GARLIC. My interest in this subject began some time back when a friend, who is totally removed from the medical profession, suggested a serious look at the curative powers of these pungent plants. Next there was the report of garlic and beer being consumed in a Tongan club with the comment, "It clears the blood." A conversation with another traveler uncovered the fact that peasants in southern France also speak of the same bulb as cleansing the blood. Such diverse sources triggered the imagination, and the search for facts was on, beginning with the Linnaean name of the reputed superherb, *Allium sativum*.

Three and a half years of *Index Medicus* revealed 32 relevant entries from the world's medical journals. A few statistics set the story. Eleven of the papers dealt with atherosclerosis, eight focused on coronary heart disease and fibrinolysis, six reported on antibiotic action, and the rest covered a variety of other clinical uses of this food substance. The general consensus of these studies is that garlic lowers blood cholesterol, inhibits atherosclerosis, and increases fibrinolytic activity in patients who have had myocardial infarction.

Since there are few known harmful side effects of garlic, except a possible loss of friends, it seems strange that a substance reported to have beneficial results with severe and common diseases would not be prescribed more frequently by doctors in the United States and Western Europe. Here another set of statistics is quite revealing. The papers cited above come from 20 different journals, most of which are not regularly read in the major medical centers. The clear exceptions are *Atherosclerosis* (three papers) and *Lancet* (four entries, all letters to the editor). The articles in *Atherosclerosis* represent work done at the Ravindra Nath Tagore Medical College at Udaipur, India, and the University of Benghazi in Libya. Twelve of the reports are in Indian journals, two in Chinese periodicals, two in Polish publications, and one comes from Japan (reporting on work done in India).

A rather fascinating chapter in the sociology of science begins to emerge from this fragmentary survey of the literature. A small, highly active group of researchers centered in India have vigorously pursued studies on garlic (also on onions) as a major therapeutic substance in treating atherosclerosis, ischemic heart disease, and hypertension. Their investigations, although carried out by standard techniques of animal physiology, controlled clinical studies, and epidemiology, are not fully believed by the world's medical authorities or else they would have been repeated immediately. It is characteristic of biological and biomedical research that one evaluates work in terms of knowing the investigators. This reflects the infinite possibilities of error, misinterpretations, and statistical fluctuations. Doubts, however, also create fraternities from which outsiders are excluded without regard to the character of their work. One thinks of an obscure Augustinian monk counting peas at the monastery at Brno.

Moving from peas back to garlic, there are additional reasons why Allium is suspect. Aficionados of the herb proclaim its efficacy in gastrointestinal disorders, hypertension, fungus infections, arteriosclerosis, leprosy, tuberculosis, the common cold, cancer, anemia, diabetes, and hypoglycemia, to mention just a few of the reported pathologies. With a sure knowledge that all of these claims cannot be true, it is easy to move to the presumed conclusion that none of them is true. This is a syllogistic fallacy that doubtless has a long Latin name, but the fact remains that it forms a subconscious part of our thinking.

Even after understanding the various logical and sociological issues, it seems very strange that the literature does not record a single U.S. study of garlic and heart disease. We are considering, on one hand, leading causes of death and, on the other, a common substance that is innocuous. If the reports in the Indian literature have any validity whatsoever, the public health potential is enormous. It certainly seems that at the very least a few rabbit experiments are in order.

But the question occurs, who would carry out the research under discussion? No industry other than garlic growers would have a vested interest in studies of this nature, so private funding seems unlikely. It is difficult to guess how federal peer review panels would react. In garlic research there are no peers in the United States. One feels that the committee would be very nervous about a proposal in this area. Something about it would smell strange to them.

Looking back over the history of medicine, *Allium sativum* is frequently cited as an herb of pharmacological importance. These applications are reviewed in *The Book of Garlic*, a contemporary compendium on that subject by Lloyd J. Harris. The Ebers Codex, a papyrus manuscript from sixteenth century B.C. Egypt, includes the strong-smelling herb in 22 of its 800 medicinal preparations. This tradition persisted centuries later in the writings of Hippocrates and Aristotle. *The Greek Herbal of Dioscorides* from the first century A.D. contains garlic formulations for a wide variety of diseases. The Romans were no less impressed than the Greeks. The famed physician Galen was aware of garlic's uses, and the naturalist Pliny catalogs some 61 remedies containing what he clearly regarded as a very powerful plant product.

As might be expected from the modern interest, India has an ancient collection of Sanskrit manuscripts regarding the use of garlic for maintaining health and curing disease. An equally long-standing history can be traced in China. Thus, the medical use has permeated the civilizations of most of Europe and Asia. The very variety of therapeutic applications raises questions as to how such a practice has been so widespread and persistent and yet has remained so enigmatic, even in the light of modern biochemical knowledge.

I, of course, retain my position as a sceptical member of the Western biomedical establishment. I'm not going to be taken in by folk remedies or reports of *Allium sativum* as an aphrodisiac. I'm not going to give up my Food and Drug Administration–approved medicines for some smelly herb of dubious value. But just in case there are some readers out there less strong willed than I and less committed to seeking firm epistemological correlations, I would like to give you my mother's recipe for delicious calf's-foot jelly with garlic:

1 calf's foot (cut in quarters)
8 cloves of garlic (coarsely chopped)
1 bay leaf
1 onion (coarsely chopped)
2 hard-boiled eggs (sliced)
Salt and pepper to taste (watch the sodium—H.J.M.)

Wash the calf's foot thoroughly, then soak in cold water for one hour and drain. Place the calf's foot in a saucepan, cover with water, and simmer for two hours. Add the onion, garlic, and bay leaf, and simmer for one additional hour. Remove the meat from the bones and arrange in a flat pan with the hard-boiled eggs. Pour the liquid into the pan and refrigerate until a hard gel forms. Cut into squares to serve.

Jurisgenic Disease

It is not clear whether *iatrogenic disease*, "illness caused by physicians," has yet become a household term, but it has certainly found its way into the courts by way of malpractice suits. Those same legal institutions are giving rise to another class of pathology: *jurisgenic disease*, "illness caused by law or lawyers." The term *jurisgenic* is derived from the Latin *ius, iuris*, meaning "law," and the condition afflicts hundreds of thousands of people in the United States every year.

The syndrome we are describing typically occurs in cases of posttraumatic pain and disability. Since a large fraction of accidental injuries either happen at work or are due to negligence or other possible grounds for liability action, recovering patients soon begin to file claims for damages. Given the adversary nature of our legal system and the litigious character of our society, such individuals soon find themselves under the care of lawyers, as well as doctors. The lawyer's professionally oriented goal is to maximize the monetary settlement, both in the interest of the victim and in consideration of his own substantial percentage of the payment. The attorney realistically recognizes that the more pain and disability his client experiences and demonstrates, the more cash there will be forthcoming for both of them.

At this point we should note we are not discussing the deliberately dishonest attorney or the malingering "victim." We are dealing with the average citizen represented by a lawyer of median honesty, trying to do his best. Nevertheless, the legal advisor in many subtle ways encourages the patient to maximize the intensity of the symptoms. The injured party quickly gets the not too deeply concealed message and begins to respond by actually feeling the increased effects or failing to respond to treatment. Both attorney and client are unaware of the seriousness of the game they play, for mind and body are so totally interrelated that thoughts of illness cannot be separated from the illness itself. In short, the patient is quickly a victim of jurisgenic disease and suffers far more pain and disability than would be expected from the degree of injury. Since the judicial system grinds exceedingly slow, these legally reinforced symptoms may persist for many years and may even become permanent.

From the patient's point of view jurisgenic disease may be extremely damaging psychologically. He wants to get better for all the right reasons yet wants also to maximize the financial return. By this time the "case" becomes a big part of his life, frequently almost an obsession. The patient may refuse what would have been a reasonable settlement because the injury is now so significant in his mind that the amount offered seems too small. The resulting situation thus may cause considerable personality degeneration.

Without being called by its proper name, jurisgenic illness is often identified by physical medicine specialists, neurologists, pain clinics, and general practitioners. These professionals may advise the patient to settle the case as soon as possible and then get down to the business of rehabilitation. Frequently that settlement is a necessary first step to recovery.

A patient with persistent posttraumatic pain and disability more severe than indicated by the injury may be referred to a psychologist or a medical hypnotist. These professionals undertake the task of aiding the individual in restoring self-image and a sense of wellness. In short, they must attempt to reverse the damage inflicted by the legal process. Medical personnel working in this area report reasonable success in behavior modification once the nature of the problem is understood.

Jurisgenic disease need not originate from lawyers. Social workers trying to protect their clients may, for example, deliver the same "don't get well" message. Veterans Administration counselors, Social Security administrators, and a wide range of other civil servants and institutional officials may all contribute to convincing an individual that pain and disability are better than health. Friends and relatives may also add to the problem with well-intentioned advice. If suffering puts bread on the table, the message is a difficult one to resist.

Recognition of this class of problems leads to the question of why it has taken so long to identify jurisgenic disease when the rarer iatrogenic disease has been such a center of focus. There are a number of probable reasons for this relative lack of awareness:

1. Practicing physicians have enough problems with the legal profession and are hesitant about initiating more interaction with attorneys. Physicians are traditionally reluctant witnesses in court because of the expenditure of time and the badgering they may be subjected to. Diagnosing the disease as being due to a patient's lawyer is, to say the least, looking for trouble.

2. The individuals responsible for jurisgenic disease, the etiological agents, are popularly seen as acting in the best interests of the patients. Their positive role is easy to see; their pernicious activity is obscured in certain incompletely understood areas of psychosomatics.

3. The reductionist, mechanical, biochemical view that has dominated medicine has given too little attention to subtle mind-body interactions. Recent findings in biofeedback, medical hypnosis, and other areas of psychology are beginning to cause a shift to a more balanced view of the individual. The perception of pain is emerging as an area where physiological and psychological factors can hardly be disentangled.

Once jurisgenic disease is recognized, we see that it is a societal malady, rooted deeply in our economy, our legal institutions, and our very ways of interacting with each other. No single treatment modality can get at a condition related to the entire social structure. However, it is possible to suggest a number of simple actions that might prove useful.

If physicians are alerted to jurisgenic disease as a known and named entity, they can be on the lookout for it in more directed ways. In cases of post-traumatic pain and disability, the history can include questions about who is giving the patient legal advice and what advice is being given. The doctor can explain jurisgenic disease and provide the patient with literature describing the phenomenon and advice on how to counter its effects.

If lawyers and social workers are warned that their efforts to assist may be having negative effects of such potential severity, they can revise their strategies. These changed procedures could still protect clients' financial interests while minimizing their exposure to psychological damage. Assuming goodwill on the parts of most of the professionals involved, their sensitivity to jurisgenic disease could be a substantial step in its control.

Most important of all is public awareness of the existence and nature of the problem. Few individuals will consciously choose sickness over health, pain over well-being, or disability over freedom. If patients are knowledgeable about the consequences of their attitudes and their physical, as well as mental, state, they can make more intelligent legal and medical decisions.

Much of pain and suffering leaves us standing mute, for we lack the knowledge and understanding to treat the disease properly and cure the underlying causes. Jurisgenic disease generates an enormous amount of pain and suffering that is within our power to alleviate if we but recognize the condition and make the appropriate adjustments, opting for health over money. Some self control and policing by the legal profession would also help. But that opens another whole can of worms.

Geriatric Gobbledygook

The receiver has just been put back on the phone after a conversation with one of the wisest and most humane philosophers I have ever known. His name is withheld because he would be genuinely embarrassed by the fine things I have to say about him. It has been my unpleasant duty to deliver a negative message to my colleague and friend. A few weeks ago he had been invited to present a college seminar series in his area of specialty. In my innocence I was unaware that Yale University has a cutoff at age 70 for members of the faculty serving in any capacity. The rule is absolute—it even applies to the limited 12-week seminar series—and no individual beyond the allotted threescore and ten is permitted to teach. Having been informed of the existence of the rule and the firmness of its application, I was required to withdraw with chagrin that which had been happily proffered a few weeks previously.

Thoughts about age restrictions naturally lead to musing about the accomplishments of older people. This university would have denied playwright George Bernard Shaw the opportunity to teach theater for the last 24 years of his life. Bernard Baruch would have been unable to instruct us in finance during the last 25 years of his career. Robert Frost could not have expounded on poetry for 19 years, and for that same period Carl Sandburg would have been denied a class in literature or biography. Will Durant, who died at 96, would have been barred from the seminar room since before the birth of today's college students. Philosopher Susanne Langer would have been in retirement for the past 19 years. Why, this very day we would reject Ronald Reagan's course on government or cinema arts.

Efforts to change this policy did not meet with much sympathy. When the mandatory retirement age changed from 68 to 70, the Yale administration decided that this figure would also serve as the upper limit for all activities connected with teaching. It is not difficult to find reasons for this inflexibility. Many professors reach mandatory retirement in full vigor and high productivity and strive to stay on and exert their influence. As an independent source of power, such teachers are a threat to administrators, especially those whose professional accomplishments are overshadowed by the eminence of their older colleagues. Rather than handling each case on its merits and having the problem of saying no, it is easier to employ rigid rules. We protect some weak officials by providing general procedures for avoiding hard decisions.

Three major issues emerge from this experience: the structuring of organizations to avoid individual responsibility; the willingness to deny our young the wisdom of scholars who have spent a lifetime pursuing demanding disciplines; our attitudes about old age.

The question of responsibility is seen in the tendency, evolved during the post–World War II period, to shift that burden from identifiable managers to committees or other vague, impersonal groups. The change in policy may be associated with the era of the anti-hero in literature, drama, and film. In collective actions individuals are not fully accountable—as in the case of a firing squad, where no one knows who has the blank. The prime example of this mode of operation is seen in *the meeting*. If you phone an administrator, at least half of the time you will be told he's in a meeting. Such gatherings are devices for spreading liability. A group sits around a table and the onus is passed from hand to hand or mouth to mouth, like a contagion. It is not entirely clear why organizations have moved from private accountability to this more public assignment of responsibility. The practice counters traditional values of Western culture and pushes individualism into the background. It could be the result of the complexity of modern society, or it could just be a loss of guts.

Given this attitude, however, rules are invaluable devices for those who prefer to administer without personal risk. Regulations are always promulgated by a vague "they," on whom blame can be heaped. The administrator is an enforcer, simply playing an assigned role. If you detect shades of Adolf Eichmann's defense, you are right: his was the limiting case of unthinkingly administering rules set by higher levels in the organizational structure. Questioning at every level would seem a reasonable price to pay to avoid the potential amorality of yes-men.

The second issue may be termed the artificial generation gap. Given our very mobile society, with the nuclear family or nuclear half-family, children tend not to know their grandparents. There is thus a gap: Culture is, at most, passed on for one generation and almost never has two chances to impose its values. In cutting off young students from their teachers' teachers, the same type of one-generation structure is imposed in the transfer of intellectual traditions. Since each age sees reality in its own light, we withhold from the young a broader perspective. The staff of the seminar series under discussion has on occasion involved young, mediocre instructors, while a reservoir of fine minds went untappped.

Today's society is oriented toward disposable products. Items are consigned to the trash rather than being repaired and placed back in service. I suppose we sometimes look at people as if they had a built-in obsolescence and must be taken out of service when the warranty expires after 70 years. This is questionable from both humanistic and scientific perspectives. Treating organisms as mechanical devices is sometimes reasonable for physiological theory but does not address values. From a biological point of view, aging in *Homo sapiens* is subject to as much variation as most physiological parameters, so that a rigid classification referring solely to sidereal time is hardly a recommended procedure.

It's not necessary to be doctrinaire about this subject. Some age categories are demanded as guides for the orderly conduct of institutions. The problem arises when we rigidly extend these rules to situations that could be handled perfectly well on an individual basis. In human social organizations there is a delicate balance between anarchy and the tyranny of rules. Establishing balance requires an ongoing search for wisdom. It's a quest that, like many others, calls for the knowledge and experience of all, including our over-70-year-olds, who must not be excluded by fiat from participation in civilization's progress. Youth means believing in the future, an attitude shared by many who have watched the earth orbit the sun for 70 or more times.

During the writing of this piece I had a wonderful phone conversation with another philosopher, a 90-year-old sage who is negotiating a contract for a four-volume series. Institutions aren't stopping people like this man, and I guess they shouldn't get me down either. After all, old age is something that even the most recalcitrant administrators aspire to.

Trees and Forests

In the Talmud one finds set down in at least two places the thought: "I have learned much from my teachers, more from my colleagues than from my teachers, and from my students more than from all of them." I first encountered that quotation several years after I had begun teaching, and it led me to think about how much I was learning from my students. That idea was a pleasant one, for it made the career of teaching a potentially lifelong intellectual growth experience.

One morning a few weeks ago, after sitting and talking with a student, the wisdom cited above came forcefuliy to mind. The young man was discussing career choices, and his comments reaffirmed how much we can learn about the future from the subtle messages hidden in young people's discussions of their personal plans. Bright students must have a nose for news to decide where the action will be when they finish their training in five or 10 years. Of course students are not totally passive agents in this process, since their choices not only respond to their visions of the future but also help to determine what will happen in years to come. This is a kind of sociological uncertainty principle, stating that the future is not totally independent of what we think it will be.

What was noteworthy in this message was that it was not isolated but representative of a recurring theme that I have been hearing of late. This particular student started with a strong interest in public health and preventive medicine. He wanted his lifework to be in areas directed toward a healthier population, with lower morbidity and lower mortality. He had been planning to apply to medical school, and his undergraduate work included the required preprofessional courses. As he matured and came to understand the nature of medical training, he realized the extent to which it focused on individual interaction between physician and patient and therefore did not stress the issues that were central to his thought. He believed that he could personally bring greater good to greater numbers in more public activities than in the one-to-one relationships of traditional medicine.

Our conversation shifted to one of my favorite topics in nutrition. If, in the United States, we were to decrease the sodium in processed foods by a factor of two and at the same time convince people to go easier with the salt shaker, then, most experts agree, the extent of hypertension in the population would dramatically decrease. Death and disease associated with high blood pressure would drop substantially and a noticeable improvement would take place in national health. Thus we have possible at the prevention level the equivalent of a miracle cure for hypertension at the treatment level. What kind of trained individuals are required to bring about such an advance? The knowledge is at hand, the technical and engineering requirements are relatively trivial, yet we face an industry geared to sodium compounds and millennia of eating habits centered on the use of salt.

Given an example of the kind of problem my young friend wished to attack, the question foremost in his mind was: Should he attend medical school? The alternatives were many. He could directly enter Master of Public Health (MPH) or Master of Business Administration (MBA) programs emphasizing health care administration. He could go the route of law and politics. As a final possibility he hesitantly suggested advertising, since, he argued, progress depends on selling ideas to legislators, physicians, and the general public.

My first response was to point out the obvious: In matters pertaining to health, both public and private, one's credibility is greatly enhanced by holding an MD and a license to practice medicine. Regardless of how one feels about this statement, it is a fact of life. Holders of other degrees have made major contributions to medicine and public health, but the politics of the situation in my student's areas of interest are all on the side of physicians. We then went on to a discussion of the long period of medical training, much of it unrelated to the actual work he would eventually carry out. We agreed that the incubation time seemed excessive. I certainly lacked the wisdom to resolve his dilemma and suggested that he get some more input from two associates, one a clinician with public health experience and the second an epidemiologist who was both physician and statistician.

The student left, and I sat staring at my desk, feeling frustrated at the kind of advice I was able to offer. It was at that point that the words of the ancient Talmudists came to mind. I was learning from my students—this student and others—something about the future of medicine. In the United States there is a growing interest in health as a positive program rather than as a response to disease. This interest emerges in exercise programs, nutrition fads, clean-air and antismoking campaigns, stress-management clinics, and a wide variety of other activities. There is a growing belief that the way we live has a profound effect on our state of physiological well-being. The management of these positive health programs has largely been ignored by physicians, who, by training, are preoccupied with treating sick individuals. As a result the fields of positive health care are sometimes being led by self-proclaimed experts, gurus with little training and less hard knowledge. Now a new generation of college-educated young people want to enter the field of positive medicine, not as instant authorities, but as trained experts bearing the imprimatur of the establishment. What do we have to offer them in the way of training programs? At the moment, an MD followed by an MPH seems about the best we can suggest. One suspects that some more novel and directed program is necessary for the new wave of students.

A somewhat unflattering analogy comes to mind. A number of years ago, when a thoughtless contractor axed gouges on the sides of some trees in order to have sighting marks for a survey, I consulted some friends in the forestry school for advice. I was told to find a tree surgeon, as the faculty dealt with the health and sickness only of forests, not of individual trees. We need the medical equivalent of foresters, professionals trained in physiology but specializing in the well-being of human populations. A significant improvement in group health would represent a major step toward assuring us of sound individuals. I would hope that the training of such "foresters" would be tied to that of the physicians in order to optimize the interactions between the one and the many and to organize the approaches in the light of modern science. The development of appropriate training programs is a challenge offered to universities and medical schools.

Much Ado About Nothing

On a number of occasions lately I have observed otherwise rational, intelligent individuals extolling the virtues of homeopathy. Not wishing to be more narrow-minded than is absolutely necessary, I decided to investigate the claims of the followers of Christian Samuel Hahnemann (1755–1843). His doctrine turns out to be an unusual olio of sophisticated concepts and sheer nonsense, defying the most firmly established laws of physics. Curiously enough, in spite of the strange mixture, there were periods during the nineteenth century when homeopathy was probably the best treatment available to most sick individuals.

Hahnemann's original theory can be divided into three postulates: 1) disease is a whole-organism property or aberration from the state of health; 2) like is healed by like, which means that substances causing symptoms at high doses will cure those same disorders at low doses; 3) the effect of a remedy is inversely proportional to its concentration.

The first tenet of homeopathy—the integrated view—occurs as an accepted idea within current system-theory studies of health and disease. Earlier concepts on simple one-to-one correlations between causes and effects have proved of limited value in a large number of pathological conditions. From Virchow to Dubos, medical scientists have spoken of multiple etiologies. The organismic approaches can be completely reductionist, as in the various mathematical models of physiological systems; or they can have a much larger psychological component, as is seen in contemporary holistic medicine. Indeed, it is on the fringes of holistic medicine that one finds many present-day homeopaths. They have adopted the perceptive and farseeing part of Hahnemann's natural philosophy without critically evaluating the scientifically untenable portions of his doctrine.

The second postulate of homeopathy—like is healed by like—is often referred to by the Latin *similia similibus curantur*. The principle is presumably a generalization from experience, but little is offered by way of the exhaustive data that must support such a synthesis. Moreover, the test of falsifiability, a foundation stone of scientific epistemology, is never invoked to deal with those cases in which *similia similibus non curantur*. Homeopathy is acutely aware of the experimental method, and there are constant references to "provings," whereby healthy individuals are dosed with a drug to find out the symptoms that it will supposedly cure. Once the substances are so "proved," there is the assumption that the dilute chemical will work as predicted. Homeopaths don't require an empirical test of the general principle; that is for them a given, a dogma of the discipline. There is thus the pseudoscience of invoking experiments at one level and ignoring the necessity of extensive data at the deeper and more important level of the validity of the general postulate.

The third aspect of the theory—dilution increases efficacy—involves starting with a drug and successively diluting it by factors of 10. The number of dilutions is designated by a factor called the potency. Thus potency 3 is one part in a thousand, potency 6 is one part in a million, etc., out to potency 30, which is one part in a thousand billion billion billion, or considerably less than one molecule per dose. This postulate comes into conflict with an underlying principle of thermodynamics so fundamental that no special name has been assigned to it. This basic assumption asserts that an isolated system will come to a state of equilibrium dependent on its physical parameters—such as temperature, pressure, volume, and composition—and independent of its history, that is, independent of how it was prepared. Let us take an example. Start out with two samples: the first, a 1% solution of sodium chloride, and the second, an extract of ground-up honeybees. Sequential dilutions are made in pure water until they have reached a potency of 30. At this dilution each final sample is just water, regardless of the starting material; the odds are less than one in a million that a cubic centimeter of the final samples will contain a molecule of the starting material. At that point no physical measurement can distinguish the two final dilutions. According to the laws of physics the two samples are in every way the same. According to the laws of homeopathy they will have different effects when administered to a patient. If homeopathy is correct, either living organisms must operate independently of the laws of physics or one of the most basic assumptions of physics must be wrong. Either way, homeopathic medicine is in direct confrontation with some of the most firmly established branches of human knowledge. Given the paucity of any positive evidence, these violations of natural law place the third postulate of homeopathy in the category of scientific nonsense.

In spite of our dim assessment of homeopathic theory it must be conceded that it enjoyed considerable success in the 1800s and early 1900s and was widely sought after. I believe we would now admit that during the early and mid-1800s, before scientific medicine began to take hold, most sick individuals would have been better off under homeopathic treatment than under the care of a standard physician. The reason is that the pharmacopoeia employed by orthodox medical professionals listed a large number of highly toxic materials, while those of the homeopath were harmless, since they contained few or no active substances. Using just water or sugar tablets as their medical armamentarium, the homeopaths were unlikely to do any harm. In addition, the allopaths (physicians other than homeopaths) practiced bleeding and similar dangerous rites. Patients of both types of practitioners were able to enjoy placebo effects and psychosomatic gains and to benefit from the fact that most illnesses will resolve spontaneously. Clients of homeopaths were spared additional risks of treatments that were often harsh and dangerous.

Given the state of knowledge throughout most of the 1800s, it was often better to do nothing than to do something; and that is precisely what the homeopaths were doing: nothing. Our story is not without a moral. It is easy to look back at a past age and suggest the value of doing nothing. It is difficult in the extreme to examine our own behavior and to conclude that there are times when no action is the optimal course of action. This is true in every field of human endeavor but particularly so in medicine, where one is presented with suffering and complaining human beings. One advantage that accrues to the homeopath is never having to admit that he is doing nothing. He always presents his patient with what appears to be an active, vigorous course of medication. The fact that the medicine is as vacuous as the theory is unknown to both the healer and the healee, thus allowing the full range of psychodynamics. Lacking these advantages, the practitioner of normal scientific medicine must nevertheless be prepared, on occasion, to make the decision that doing nothing is the best course of treatment.

Why has there been a return to homeopathy in the current age of science? The answer is a familiar one. In the face of difficult questions or problems that lack solutions, individuals turn to doctrines that are without a rational basis but have a ring of conviction about them. The history of human folly is, after all, rich in examples of such ideologies. The movements arising from such doctrines offer solace to the sick and the confused. In the case of homeopathy, "nothing" has gone, and continues to go, a very long way when that nothing is accompanied by hope and faith.

Through an Ophthalmoscope Darkly

Psychologists have dealt with the concept of self-images, or "ideational constructs of the self which supply unity and striving in the world of people and things." Such internalizations can be regarded as normal or pathological, depending on their correspondence with publicly held perceptions of friends and associates. Self-images belong to professional and social groups as well as to individuals, and here again we face the relationship of internal and external views. To be philosophically complete, both sets of ideas must be tested against some reality out there that can at best be dimly perceived in dealing with such a psychological domain.

Thus the medical profession, when viewed from the inside, seems to consist of a highly selected group of cosmopolitan individuals. When these conceptions are tested against certain numerical data now available, this self-image fails to correspond, and the difference appears to be increasing each year.

The facts of the matter are well documented in the 32nd edition of the handbook *Medical School Admission Requirements*. Published by the American Association of Medical Colleges, the annual volume is the ultimate "how-to" book for medical school applicants. Within its pages lies a wealth of statistical information for those concerned with the sociology of the admissions process in the United States and Canada.

A prime item of current interest emerging from a study of the book is that approximately four out of every five (78%) entering candidates are residents of the state in which their medical school is located. There is a quantitatively clear contrast between the sophisticated international image of medical research and the somewhat provincial view seen when one examines admissions policy. Analyzing this situation in terms of the practices of individual schools reinforces this picture of local bias.

Almost half of all medical schools have entering classes consisting of 90% to 100% state residents. Of these, 10 have gone the complete route to provincialism and admit no out-of-state residents. To be truly nationally representative, I would judge that the student body should consist of less than 30% in-state residents. The number of such open institutions is small. They are Brown, Dartmouth, Duke, Georgetown, George Washington, Harvard, Howard, Johns Hopkins, Meharry, Oral Roberts, Vanderbilt, Washington University (St. Louis), and Yale. Another group of schools might be categorized as seminational, admitting more than 50% but less than 70% nonresidents. They are Boston University, Chicago-Pritzker, Creighton, Northwestern, St. Louis University, Stanford, Tufts, Tulane, and Vermont. All other institutions of medical education must be classified on a scale ranging from local to exclusionary.

Given the strong bias toward state residents, it is not surprising that less than 2% of those admitted are not citizens of the United States, and almost all of this small group have permanent-resident status. It is virtually impossible for a genuine foreigner to receive a medical education in the United States.

The reasons for the extremely provincial character of the selection process are not difficult to fathom. Medical education is extremely expensive and is largely funded by states. These governmental units spend their megabucks to assure themselves of a supply of physicians and thus provide health care for their citizens. Monies are appropriated by state legislatures. Members of such bodies are among the least mobile citizens of the country. Election campaigns for assemblies and senates are by district and tend to stress length of residence as well as local issues. Very often the composition of the entering classes in state institutions is mandated by these legislatures.

Given these legal restrictions, an exclusionary policy greatly simplifies the work of admissions committees. For example, in choosing the 1980 entering class, the Louisiana State University School of Medicine in New Orleans had to review 656 in-state applications to choose 175 candidates. The 204 completed forms from misguided nonresidents probably did not have to be read. For the same year Johns Hopkins, with its completely open policy, had to review 3,091 files to fill 120 positions.

Why does the extreme territoriality of medical education give us some feeling of unease? I suspect the primary concern is that the students miss the broadening influence of association with people of a wide range of cultural backgrounds. This will not affect skills as a practitioner but will influence attitudes that an individual brings to his practice. When more and more aspects of life are undergoing a nationalization due to mass communication, it is noteworthy to find an important sector of social activity that is becoming less cosmopolitan.

Within the pages of *Medical School Admission Requirements* one also finds other facts running counter to the professional self-image. For example, the number of places in medical schools has risen steadily from 15,269 in 1976 to 16,590 in 1981. At the same time, the total number of applicants has declined from 42,155 to 36,100. This has been a continuing trend in the past 10 years. Thus while only one out of every three applicants was admitted to some school in the early 1970s, at present about one out of every two is admitted.

Because individuals file multiple applications, sometimes as many as 30, the illusion exists, both externally and internally, that entry into the medical profession is highly selective. The Johns Hopkins figure given above shows only one out of every 25 candidates is admitted. Each of the training institutions exercises considerable selection, but the profession as a whole is now faced with a rather low degree of overall choice concerning who enters. I suppose we must face the question of why, when it is statistically easier to get admitted, do fewer students make the effort. The answer to that question must involve a changing perception of medicine by the college-age population.

The admissions data alter our image of the entering class from that of a cosmopolitan, rigorously chosen group to a collection of individuals who are both locally oriented and largely self-selected. This, I submit, will come as a blow to the collective ego of the medical profession. If the number of places in state medical schools continues to rise and the applicant pool continues to diminish, then our somewhat pejorative description of the entering class will come progressively closer to an accurate appraisal of the profession.

The fact-packed publication of the Association of American Medical Colleges is our Delphic Oracle for the future of the health professions. Trying to interpret the message should occupy the attention of those concerned with the orderly development of medicine. It's also of considerable interest for the rest of us. Carefully analyzing the numbers is an antidote to the problem of a distorted professional self-image.

Little Black Boxes

The events of which I write occurred many years ago and would be all but forgotten except that the questions they raised will not go away. The episode began innocently enough with a phone call from an attorney (assuming phone calls from attorneys can ever be innocent) asking if I could serve briefly as a consultant to a local manufacturing company that he was representing in a dispute with the Food and Drug Administration. We arranged to meet for a briefing.

Over dinner the case was presented. An electrical manufacturer had developed and marketed an air purifier that was being sold through department stores. Accompanying each unit was a warranty card with a space for comments. As these replies were read each day, a strange medical wonder story began to unfold. Asthmatics and individuals with hay fever and allergic rhinitis were, in a totally unsolicited way, hailing the unit as a remarkable therapeutic device in improving their breathing and allowing them to sleep in comfort. These comments were transcribed from the warranty cards by the advertising copyriter. Business boomed; tens of thousands of units were sold, followed by that knock on the door from the Devices Division of the Food and Drug Administration. Among other items, they wished to know the physiological mechanisms involved in treating allergic respiratory diseases. My mission, did I care to undertake it, was to measure the physical properties of the air emerging from the unit. With a slightly whimsical feeling, I left the restaurant with a brand new air purifier tucked underneath my arm.

Working in my home basement laboratory and with an air of scepticism, I disassembled the device and noted a fan, a fiberglass air filter, three small ultraviolet bulbs, and associated circuit elements. In fairly short order I convinced myself that the unit circulated room air, taking out a measurable amount of particulate matter and adding back minute quantities of ozone and a substantial current of negative ions. With little to go on, the next step was a plunge into the literature on the physiological effects of negative ions. That literature was a mess. For every statement in a journal article, a negation could be found in a publication of equal standing. There were many indications of the beneficial effects of negatively ionized air in respiratory distress; there were also clinical trials that showed no benefits at all. After noting the logical difficulty of concluding anything from contradictions, I moved on to the next step.

Being of a somewhat suspicious temperament, I asked permission to read the user comments on the warranty cards. While the theory was ambivalent and confusing, the phenomenology was overwhelming. File after file was bulging with cards extolling the "miracle cure" of respiratory ailments. Some of the cards had a religious fervor; others were poignant in their descriptions of transformed lives. Reading these cards was an unusual experience because of the intensity of emotion they exuded. It is a bit strange to pick up a commercial instrument like a warranty card and find "God bless you" scrawled across it.

In an effort to comply with the Food and Drug Administration, three procedures were followed: (1) Clinical trials were instituted; (2) physicians were identified among those who had sent in warranty cards, and questionnaires were sent out; and (3) a statistical study of purchasers' comments was undertaken. Double-blind clinical trials were complicated by the difficulty of designing ineffective control units that looked, sounded, smelled, and felt like the standard units. Substantial symptomatic relief was obtained with the devices, and smaller but still appreciable benefits were found with the controls. Many physicians replying to the questionnaire reported personally using the units, as well as advising their patients to do so. The study of the purchasers' replies made it clear that the response was widespread and did not represent small-scale statistical fluctuations.

In summary, here was a device that provided symptomatic relief for a large group of patients with complaints ranging from mildly irritating to disabling. There were no indications of any harmful side effects. The ozone concentration was orders of magnitude below toxic levels, and by all judgments the machine represented no hazards of any kind. Finally, aside from the uncertain effects of air ionization, no one could provide an adequate physiological explanation of why the unit was producing such dramatic results.

In the face of this kind of evidence the Food and Drug Administration, using the full power of a federal regulatory agency, kept constant pressure on the relatively small manufacturing company. It was a David and Goliath situation in which the giant won. The manufacturer folded, and the remains were bought out, for whatever reason, by a large corporation.

As a bystander watching the corporate carnage, I kept wondering if the demise of this company and the disappearance of this air purifier really served the best interests of the American people. Should the public be kept from purchasing a device that is harmless and strikingly effective in a large number of cases and whose mechanism of action is not understood?

I had entered this situation as an unquestioning friend of legislation governing medical devices. It is clearly a method of protecting the public from quacks of all sorts selling innumerable gadgets like pyramids, orgone generators, and all manner of little black boxes that rob the user of hard-earned dollars without giving anything in return. The argument is also offered that the availability of such devices keeps people from seeking medical consultation, thereby leading to a deterioration of their health. In all such legislation the question is how far society should go in shielding us from our own folly before that protection becomes oppressive. We are constantly confronted with the legitimate role of a nation in safeguarding its population versus "Big Brother," who knows what is best in all intimate aspects of our lives. The case in point, in its full detail, emerged in my mind as an example of Big Brotherism. The moral, of course, is that anyone given regulatory power will commit excesses, and a system of checks and balances is required to constrain those abuses. The cost of litigation with the federal government is very great, and only large corporations are able to carry it off successfully, leaving few approaches for smaller companies dealing with the bureaucracy. They usually go out of business in this situation.

By the criteria set by the Devices Division few over-the-counter items would pass the test. Consider the adhesive bandage strips commonly applied to minor cuts. How would you design a double-blind test with a noneffective strip that looks and feels just like the real thing? What is the mechanism of action of these devices? Do adhesive strips keep people from going to the doctor for cuts that require medical attention? And if these bandages are dangerous, what about the potential hazards of adhesive butterflies? Has the whole concept gone too far? The FDA should protect us from dangerous devices, and the FTC should protect us from fraud. Beyond that, it is hard to see why the federal government should be concerned with the ontology of causal relations and the conceptual problems in constructing mechanistic models.

A second more philosophical issue emerged. The Food and Drug Administration appears to have an operational view of science that is naive in the extreme. Cause and effect, treatment and cure, are viewed from the point of view of an elementary textbook, ignoring the fact that in the practice of medicine such sharp distinctions are rarely possible. We seldom know the detailed mechanistic events that intervene between treatment and the results of that treatment. The criteria used in evaluating new devices and drugs would, if universally applied, bring medicine to a standstill. There are deep methodological issues in evaluating treatments, and some extend to the epistemological basis of dealing with very complex, partially understood systems like the human body. It seems arrogant for the regulators to assume that in their simplistic approach they have all the answers. By so doing, they may harm as well as help their constituencies.

The Ecosystem Within

A general awareness has recently emerged that we are all part of an ecosystem, a large network of interacting flora and fauna whose long- and short-term interactions play an essential part in determining our natural surroundings. Less obvious to the eye is the inner ecosystem, those hundreds of species that live within us and upon us and occupy an equally important role in our lives and our well-being. Each of us acts as host to billions of bacteria, yeast, fungi, viruses, protozoa, and other assorted "wee beasties" that coexist with our person. In short, as well as being individual *Homo sapiens*, each of us is an ecological system in all the biological meanings implied by that term.

To bring the matter into sharper focus, consider that a human adult is made up of 10 trillion cells, each derived from a single fertilized ovum. Dwelling within the mouth and gastrointestinal tract, upon the skin, and in various other recesses, are 10 times that number of alien microorganisms. Thus 90% of the cells that I carry around with me as I walk the streets are not mine, either genetically or by developmental lineage. They have come to me from the outside world and have been fruitful and multiplied. Since they are my closest biological associates, it is important that I know a little more about them.

As a fetus lies in the womb just prior to birth, it is uncontaminated from a microbiological point of view and thus entirely free of any germs, with the possible exception of a few stray viruses that may have crossed the placental barrier from the maternal blood supply. If such an individual were removed by cesarean section under absolutely sterile conditions and reared in a completely uncontaminated environment, he could live an entire life free of microbial associations. Although this has not been practical for humans, such experiments have been carried out with mice, rats, chickens, and dogs, and much of the knowledge of the effects of human microbes comes from a study of these germ-free creatures. Many experimental animals have spent their entire lives in such a gnotobiotic state, that is, without physical contact with other organisms.

For most humans the first contact with outsiders occurs during the process of delivery, when the skin, nose, mouth, and eyes all become contaminated with organisms from the genital passage of the mother. As we begin to drink and eat, more germs enter our alimentary canal and find their way into niches where they perform their major function, replication. As we age, successively more microorganisms become associated with our persons until at maturity some 500 or more species have become lifelong companions.

From the point of view of ecology the increase and changes of these microbial populations have a striking analogy to a process that has been designated *succession*: the growth, development, and distribution in time of the biota within a habitat. As an example of ecological sequence, consider a fresh volcanic flow on a moist tropical island. As the hot lava begins to cool, it is sterile, devoid of life. The wind blows in blue-green algae and lichens that begin to grow on the hardened basalt. Microbial action, along with weathering, gives rise to soil within which insects, algae, and moss may grow. Seeds take root and ferns sprout, followed by trees. Each kind of organism interacts with others and alters the environment. As the system ages, some of the pioneer species may no longer be found. Finally a tropical forest stands on what was once a barren waste. If the climate changes or the environment is otherwise altered, a further succession will take place, leading to a population of organisms appropriate to the new conditions.

In the newborn infant the intestinal tract is as sterile as a fresh lava flow. By the time maturity is reached, hundreds of different species of microorganisms will be found in the colon. The time from birth to maturity must witness a succession almost as rich and variable as that of the tropical forest, though totally hidden from view. Zones characterized by differences in pH, oxygen pressure, and concentration of nutrients develop in the intestines. These habitats lead to ecological niches characterized by the species that occupy them and the relations between these species.

There are human diseases that may be described as ecological catastrophes of the gastrointestinal flora. Some cases of "turista" fit into this category. This syndrome may be induced by large bacterial population shifts due to the introduction of new species and varieties. Other examples are the upsets in bowel pattern following antibiotic therapy, which selectively kills sensitive colon strains, leaving unbalanced populations to competitively establish a new steady state. The rumbles in our guts are echoes of the Darwinian struggle for survival being waged within.

An early successional disease is infant botulism. In adults the occasional appearance of a *Clostridium botulinum* spore in the large intestine causes no difficulty because the competition from other bacteria keeps the strain in check at a low population level. In very young infants the ecosystem may be underdeveloped in competing species, allowing the invading pathogen to undergo a growth surge. The toxin produced by the metabolizing Clostridium will enter the bloodstream and, depending on the amount absorbed, will cause severe illness.

The colon as a habitat or series of habitats is dependent on the food we eat. This is dramatically seen in the distribution of *Sarcina ventriculi*. In the feces of vegetarians there may be as many as 100 million of these bacteria per gram. Among meat eaters the species is almost never found.

A fascinating case of succession is the formation of a dental cavity. The first event is the appearance of plaque on a tooth. The deposited material comes from the saliva and may contain molecules produced by oral bacteria. Pioneer species consisting of a group of streptococci led by *S. mutans* then take hold and begin to multiply on the plaque. As they grow, they are joined by actinomycetes and unclassified filamentous bacteria. The acid produced by the microbes begins to etch the tooth, and a low-pH environment is created where lactobacilli may flourish. As the hole begins to develop, anaerobic habitats are produced where Veillonella take over. Finally, in the mature cavity a number of bacilluslike rods are found. The entire process is a paradigm of an ecologist's description of time sequence in a developing habitat.

There is a whole new feeling that comes from regarding oneself as an ecological system. After long years of learning to hate microorganisms of disease, how are we going to relate to all those creatures for whom we are Mother Earth herself? We cannot ignore them, for the metabolic activity of the gut bacteria is potentially equal to that of the liver. They are our intimate associates, yet they are genetic strangers separated by billions of years of evolution. They participate in our vital processes, yet gnotobiotics tells us that we could get along without them. They are parts of our body, yet they are blocked from our brain and thought processes. They seem like mere chemical reagents, yet they share with us the cellular features of life and the most basic molecular mechanisms. I don't know whether it is humbling or ennobling to be a collection of habitats for these creatures. At the very least it makes us acutely aware of the interrelatedness of diverse living organisms.

SECTION THREE

THE PACE OF LIFE

Mind over Matter

The Pace of Life

Every now and again a paper appears in the scientific literature that strikes us because of its simplicity and relevance to everyday thoughts. Such a journal article may set a stream of consciousness flowing in unpredictable directions. A case in point is found in the February 19, 1976, issue of *Nature,* and I still find myself from time to time musing about its meaning. The study, "The Pace of Life," was carried out by psychologists M. H. and H. G. Bornstein. In each of 15 places around the world the authors "sampled and measured the tempo at which local inhabitants walked the main streets of their cities and towns." They unobtrusively observed individuals walking "alone and unencumbered" (as if any of us walks through life unencumbered).

In Czechoslovakia, France, Germany, Greece, Israel, and the United States the investigators watched walkers on "dry sunny days of moderate temperatures." And, to my great surprise, these lone walkers around the world follow a mathematical relationship between the speed at which they walk and the population of the city or town they are strolling in. The precise form of the equation states that the average speed of walkers (in feet per second) is equal to a constant, 0.05, plus 0.86 times the logarithm of the population ($V = 0.05 + 0.86 \log P$).

The data are given for areas with populations ranging from 365 to 2,602,000 and where average walking speeds are between 2.2 and 5.9 feet per second. A mathematical relation like this one invites us to consider its meaning as well as to extrapolate to domains not covered in the original study. So we predict that Robinson Crusoe and his man Friday living alone together on an island would have moved around at an average speed of 0.31 feet per second. Even more surprising, we calculate that Henry David Thoreau, living at Walden Pond without companions, would have crept about at a snail's pace of 0.05 feet per second. Little wonder that he had time to make such detailed observations of the woodsy surroundings.

Of course, the raw data do not exhaust all the human aspects of walking speed, and we can allow our imagination to fill in the details. For example, the first point on the graph is based on the tiny town of Psychro on the island of Crete. With a population of only 365, the inhabitants were leisurely strolling at about 2.8 feet per second. But the mathematics tells us nothing about how the birth of a baby or demise of a patriarch would affect the pace of life in this town.

Safed, Israel, has 14,000 residents, who perambulate at just under 4 feet per second. It is perhaps difficult for them to move too fast on streets haunted by the thoughts of the Kabbalists and mystics who have been living here for over 500 years. Are the heartbeat and footsteps of Safed influenced by memories of the many times that this area has changed hands during the Crusades or the rebellion against Rome? Just another data point, of course, except that mystics talking to God may walk at a different pace than the rest of us.

The Corsican city of Bastia has a population of about 50,000 people, who have an average walking speed of almost 5 feet per second. The Bastia value is considerably higher than we would expect from the mathematical formula. These Corsicans seem too animated for the size of their town. Are they inspired by Napoleonic memories, or are they perhaps propelled onward by the celebrated wines of Cap Corse? A single value on a graph leaves much to the eye of the beholder.

Next comes Iráklion, with its 78,200 inhabitants walking somewhat more slowly than would be predicted. Do these people of Crete spend time thinking of their Minoan ancestors? Do they tread on the ground carefully, remembering the great earthquakes of 1664, 1856, and 1926, which have destroyed many of the famous architectural works of antiquity? Do these olive oil and leather merchants recall the German invasion of 1941, with damage from bombings as severe as that experienced from the earthquakes? Statistical information reveals little about the men and women walking the streets of this Greek island.

There is a more personal note for me in the reported data on New Haven, Connecticut, with 138,000 inhabitants. Was I one of the 20 people who were observed alone and unencumbered, walking at 4.3 feet per second between two marks 50 feet apart? I spend a good deal of time walking along these streets. Sometimes I am planning an experiment, and I think I then slow down to a pace even slower than the good residents of Psychro. On other occasions there is the problem of getting to a lecture on time, and I suspect I rush about at 8 feet per second or more.

The standard deviations of these data are also of interest; they tell us how much variation there is in the speed at which people are walking. On Flatbush Avenue in Brooklyn most people were clocked walking within ± 10% of the average speed. This is rather a more disciplined life than one usually associates with a locale that is the butt of so many jokes. The town of Dimona, Israel, on the other hand, is reported to have a variation of ± 70%. Is the value a typographical error, or are the residents of the Negev really both runners and crawlers? Could the presence of a nearby nuclear reactor both speed up and slow down the comings and goings of these inhabitants of Dimona? As we say in our progress reports, more research is needed.

The fastest walkers in the entire study were observed in Wenceslaus Square in Prague. These urban Czechs were moving about at 5.88 feet per second, which figures out to be 4.01 miles per hour. This pace is considerably speedier than would be expected for a population of 1,092,759 souls. From the size of the city alone the prediction is 3.57 miles per hour. One wonders how fast these people might move were they not hearing the tread of Soviet footsteps behind them.

The study's slowest walkers come from Itea in Greece, where the inhabitants keep to a leisurely 2.27 feet per second, or 1.55 miles per hour. I know nothing of Itea, but some day, when the pace of life in New Haven weighs heavily on me, I should like to visit this town of 2,500 and walk slowly back and forth on the main street, allowing my frayed nerves to be soothed by the Itean metronome, which proceeds at a rate less than 40% as fast as that of Prague.

The *Nature* article does suggest an amateur scientific activity that can enliven anyone's travels. Along with your *Fodor's Guide*, pack a 50-foot tape, a piece of chalk, and a stopwatch. Whenever you come to a new town or city, place two chalk marks 50 feet apart somewhere on a main street and unobtrusively time people walking between the two spots. In police states I suggest that you first clear your activity with the local constabulary and assure them of the innocence of your activity. Avoid red-light districts, such as those in Amsterdam and Hamburg, lest your activities call forth a hostile response. In Moscow simply don't try the experiment. However, whenever you are bored or your spouse wants to spend some time shoping for items that are not of interest to you, simply whip out your Pace of Life Kit and start measuring. Try to clock at least 20 people at each location. Send your results to me, and I will keep a large graph on which everyone's reports will be entered. When you have sent me sufficient data, I will naturally submit a paper to *Nature*. You will have contributed to knowledge, and I will have produced another publication, all of which seems fair enough.

The Time of Your Life

Although many tourists specialize in museums and cathedrals, I am more frequently drawn to bistros, breweries, and universities. The lure of the checkered tablecloths seems so close to the lore of the contemporary inhabitants that it's sometimes difficult to balance past and present when visiting a new land. Besides, there's always the hope that some flamboyant artist will come in and paint a picture on the tablecloth. That explains the bistros. The inclusion of breweries clearly stems from a deep interest in microbiology (wineries might substitute), and universities are, after all, a bread-and-butter matter to some of us.

Given such idiosyncratic interests, a recent trip was atypically a two-church tour. First we viewed the 370-foot-tall tower of the Cathedral of Utrecht, a structure that has loomed over that city since the 14th century. Now we stand in King's College Chapel, Cambridge, staring up at the vaulted roof far overhead. The magnificence of the edifice does indeed match Wordsworth's lines:

> *These lofty pillars, spread that branching roof*
> *Self-poised, and scooped into ten thousand cells*
> *Where light and shade repose, where music dwells.*

In getting information about King's, we learn that the first stone was laid in 1446 and the building was completed in 1544. Thinking about those 98 years sets off a resonant chord because of a recent conversation about the several centuries required for the completion of the Utrecht Cathedral. This recalls Herman Melville's comment that the "great Cathedral of Cologne was left, with the crane still standing upon the top of the uncompleted tower. For small structures may be finished by their first architects; grand ones, true ones, ever leave the copestone to posterity." These houses of worship that were so long in the building suggest a clear distinction between the current age and the medieval and Renaissance periods. It is doubtful that anyone today would begin a structure whose completion date was more than 5 to 10 years in the future. We live too much in the present to be able to relate to the consequences an action will have 100 years in the future. Few fathers nowadays plant olive trees so that their son might harvest the first crop. Actually, not too many fathers stay around to watch their sons grow up, let alone tend their olive trees.

King's Chapel revives some ideas that had been engendered a few years back during another bit of tourism, wandering through the British Museum looking for the Rosetta stone. There is something very comforting about deciphered mysteries. The Egyptology section's superb collection of sarcophagi and funerary art had triggered queries about the ancient Nile civilization. Those questions, which had lain fallow for a time, were being partially answered in Utrecht and Cambridge and in all those other locales where the rising spires of great stone structures signaled a former age's religious commitment. The thoughts that had been so perplexing while I was wandering through the artifacts of ancient Egypt led to the question of why so much of that civilization's wealth, art, and technology had gone into death. How were we to understand the obsession with death that had consumed much of that society's output, starting with the first dynasty and continuing for millennia. The tombs were filled with celebrations of eternal life, even though the very construction and filling of those tombs must have imposed a heavy burden on the day-to-day life—the artistic and engineering productivity—of an entire nation.

The common theme that runs through the activities of dynastic Egypt and preindustrial Europe is a contract with immortality. The ancient Egyptians wanted to live forever, to watch uncounted floodings of the Nile and its regeneration of the land, and to enjoy the pleasures of life. Some theoretical genius or geniuses thousands of years ago postulated the concept of life after death, and there arose a highly structured set of rules demonstrating how the quality of that afterlife could be guaranteed and its physical state enhanced by the contents of the tomb. One wonders how the adherents could have been so certain that their theoretical ideas were right, but certain they were, as is most tangibly demonstrated in the wealth of artifacts that have been transported from the crypts of the Nile Valley to the museums of the world.

In looking at the beliefs of any group, one must constantly face the realization that ideas that appear strange and bizarre to us must have made perfect sense within that other society. How else can we explain these people living their lives and expending enormous efforts within the framework of their beliefs. It is nevertheless very difficult to develop empathy with ideas that seem so dubious within our scheme of things. A corollary states that our most cherished beliefs will look equally bizarre to other people in other times.

For the builders of King's College Chapel and the vast number of other churches of Europe the contract with immortality had changed. The concept of life after death persisted, but its character had altered to the more abstract view of survival of the soul. If Dante portrayed a vivid heaven and hell, these were but symbolic representations of something that was nonmaterial. Tombs were no longer filled with food and clothing and equipment for afterlife. Happiness after death no longer depended on the physical state but upon a relationship with an all-powerful and unseen God. One of the chief ways Western society developed for maintaining a good relationship with the Lord was building magnificent structures where He could be worshipped. By prayers offered within these buildings could the "faithful hearts . . . livest . . . with the Father . . . for ever and ever" (from a prayer for King's Chapel).

From the perspective of eternal life the hundred or so years required to build a cathedral seem reasonable enough. Each stage of construction renews the contract for perpetual care; each city and college requires a large and noble material symbol of its ties to the Almighty.

Reviewing 5,000 years of contracts with immortality very naturally leads us to inquire about what kind of deal our present society is making. The answer is made difficult by the plurality of contemporary thought and the blurring of vision that comes from looking at events too closely. Nonetheless, some generalizations suggest themselves. We have, on the whole, lost the confidence in afterlife that so characterized the other civilizations we've been looking at. Not wishing to let go entirely, we have substituted the less abstract dream of longevity for the uncertainties of the more highly prized immortality. We are thus willing to commit a very substantial and ever increasing fraction of gross national product to medical research and health care and a much smaller percentage to religious institutions. Medical centers have replaced cathedrals as foci of our lives, and doctors have replaced priests as the expounders of the mysteries that seem most important. Longevity has replaced immortality as a central theme, and length of life often becomes a fetish obscuring quality of life. Whether we cure cancer becomes, for some, more important than our compassion for each other.

When Tel Bethesda (site of the National Institutes of Health) is excavated 3,000 years from now and the CAT scanners and cobalt sources and dialysis machines are put on display, one wonders how the museum-goers will look back on us. Will they say that we made a religion of sickness as the Egyptians had made a religion of death? Will they find our equipment as bizarre as the great carved sarcophagi and gold death masks? Will they ask why so much of the productivity and technology of our society went into being sick in such elaborate and ritualized ways?

These unanswered questions disappear into the vastness of the Great Vault. Light pours through the Great Windows and illuminates the work of uncounted masons, woodworkers, glaziers, and craftsmen of every description. We cannot know whether the King's College contract with immortality was a valid one; we know for certain that it inspired the "Architect who planned . . . this immense . . . And glorious Work of fine intelligence."

Omar and the Ayatollahs

Reports of the faithful smashing wine bottles in the hotels of Tehran created a sense of déjà vu. The quatrains of an early Persian kept coming to mind, recalling previous breakings of jugs. Nine hundred years ago the poet Omar inveighed against the religious orthodoxy that is currently government policy in Iran. The great versifier loved wine, women, and song and opposed those who would brand his desires as sin. The Imams of his time and the Ayatollahs of today reject wine and song and prefer to publicly view women—eyes only. Regarding the spilling of the fruit of the vine, Khayyám responded with the lines

Why, be this Juice the growth of God, who dare
Blaspheme the twisted tendril as a Snare?
A Blessing, we should use it, should we not?
And if a Curse—why, then. Who set it there?

These verses, so eloquently translated by Edward FitzGerald, take us back to A.D. 1030, the date of the death of Sultan Mahmud, the ruler of Persia. Although we are uncertain about the exact date, the same year may well have witnessed the birth of the celebated Ghiyath-ud-din Abul Fath 'Umar Khayyami [Omar Khayyám].

Into this Universe, and Why *not knowing*
Nor Whence, *like Water willy-nilly flowing;*
And out of it, as Wind along the Waste,
I know not Whither, *willy-nilly blowing.*

Four years later the Turkish tribe of Seljuks revolted against Mahmud's successor, and by 1038 they had captured the city of Naishápúr, where the young Omar lived with his family. A new dynasty was established.

Whether at Naishápúr or Babylon,
Whether the cup with sweet or bitter run,
The Wine of Life keeps oozing dorp by drop,
The Leaves of Life keep falling one by one.

After its capture the province of Khorasan, in which Naishápúr was the capital, remained tranquil enough for the young scholar to pursue his studies in metaphysics, mathematics, and astronomy. His city boasted six colleges and a great astronomical observatory.

Myself when young did eagerly frequent
Doctor and Saint, and heard great argument
About it and about: but evermore
Came out by the same door where in I went..

When he matured, Omar wrote nine major works in the natural sciences, mathematics, and metaphysics and a tenth volume of poetry that was collected posthumously. Only three of his books survive: a monograph on some difficulties in Euclid's geometry, a treatise on algebra dealing with cubic equations, and a book of verse for which he has best been known through the ages.

Ah, but my Computations, People say,
Reduced the Year to better reckoning?—Nay,
'Twas only striking from the Calendar
Unborn To-morrow, and dead Yesterday.

The 11th century was characterized by the mutual influence of Arabic Islamic culture and Persian thought. Persian scholars were instrumental in the revival of Greek philosophy among the Muslims. The province of Khorasan had Buddhists, Vedantists, and Zoroastrians among the competing schools of religious belief. Omar, however, showed little patience with matters theological, believing that such things were unknowable.

Why, all the Saints and Sages who discussed
Of the Two Worlds so wisely—they are thrust
Like foolish Prophets forth; their Words to Scorn
Are scattered, and their Mouths are stopt with Dust.

The world of Khayyám embodied many contradictions. The Shiah faith was becoming a national religion, and its leaders, the Imams, were acquiring influence. The mystical brotherhood of Sufis also was gaining a following. There was great ferment in the Islamic world. Yet the ordinary Persian, outwardly given to religious devotion, did not easily renounce his weakness for wine. The tavern and mosque at times coexisted.

Before the phantom of False morning died,
Methought a Voice within the Tavern cried,
"When all the Temple is prepared within.
Why nods the drowsy Worshiper outside?"

During the first half of Omar's life Naishápúr was sympathetic to the free thought of the followers of the scholar Avicenna. Then a reactionary period of religious orthodoxy set in. In the West, the world of Islam was beset with the Crusades, and a religious fury inflamed much of civilization. The last 30 years of Omar's existence were in a period dominated by fanaticism, superstition, disorder, and barbarism.

The Worldly Hope men set their Hearts upon
Turns Ashes—or it prospers; and anon,
Like Snow upon the Desert's dusty Face,
Lighting a little hour or two—is gone.

The intellectual and freethinker of Naishápúr turned inward and composed quatrains as an expression of his secret thoughts. They were not set in manuscript form in his lifetime but were known from oral tradition, perhaps recited to his friends.

Waste not your Hour, nor in the vain pursuit
Of This and That endeavor and dispute;
Better be jocund with the fruitful Grape
Than sadden after none, or bitter, Fruit.

I wonder if there are those in Tehran or Naishápúr today who pull down their shades and take out a worn copy of Omar and read about "a Book of Verses, . . . a Jug of Wine, . . . and Thou." Is some freethinking professor quietly composing quatrains in secret to be transmitted only to his most trusted friends?

And has not such a Story from of Old
Down Man's successive generations rolled
Of such a clod of saturated Earth
Cast by the Maker into Human mold?

The Paradox of Paradoxes

We sat across the coffee table looking uncomprehendingly at each other. His eyes, which I had remembered as dancing some eight years ago, now appeared deeper set and peered more intently. The seemingly carefree graduate student of science had matured into a follower of Eastern philosophy who devoted himself rigorously to meditation and the search for universal truth. We had been talking for most of the night, so that both impasse and exhaustion were upon us.

The former student, recognizing a certain sympathy for metaphysical thought on the part of his teacher, had returned to share his newfound enlightenment and to seek my participation. In glowing and excited terms he repeatedly spoke of experiencing a higher consciousness, which could only be existentially known and was beyond our ability to verbalize. I expressed doubts about experiential phenomena that cannot be communicated and therefore require a great commitment of time, effort, and freedom before they can be minimally evaluated. There was a keen disappointment on the part of my fellow seeker, who felt as if he were offering a gift that was being rejected.

Some while later I found myself in the foyer of a Hare Krishna temple discussing profound matters with the young daughter of family friends. Her apple-pie manner and middle-American appearance strangely contrasted with the background of Vedantic mysticism surrounding us. I had come to convey greetings from her parents in a distant city but somehow could not resist inquiry into the philosophical roots of her newfound beliefs. She admitted that her knowledge was elementary and answered only in the words of her teachers. My questions and her replies made little or no contact, for too wide a gap existed in the very vocabulary and syntax. It is supremely difficult to start with the devotion of the *Bhagavadgītā* and the skepticism of David Hume and arrive at anything like a meaningful synthesis. When I continued the dialogue by mail, her single response consisted of published tracts with the appropriate sections circled. Her answers, which may have made perfect sense from within, were couched in terms that said little or nothing to an outsider.

On a subsequent occasion, while wandering through the Me'a She'arim, the Orthodox Jewish section of Jerusalem, I chanced into a small house of prayer and found two Hasidim deep in the study of the Talmud. These young men, dressed in black, wore beards and earlocks and by their appearance could have stepped out of 18th century Poland. We fell into a conversation, and to my great surprise, I found that they had both come from the United States and had family backgrounds basically secular in outlook. They spoke of their happiness in having rediscovered their ancestral faith, and their behavior gave no reason to doubt that the joy of piety was real. Again my attempts to explore the chain of reasoning that had brought them to that point resulted in a minimum of communication. To these two, wrapped within the protective cover of their prayer shawls, my epistemological doubts had little meaning. For these believers conversing with me, the answers existed in full in the books that were spread out before them.

You may think that my life is spent inquiring of the young in imitation of Socrates, and in part a mea culpa must be enunciated. But I cannot resist the hope of one day gaining some insights that will be operative within my framework of nondogmatic rationalism. All these past meetings serve to introduce the next encounter.

On the lawn in front of the biology building, a young man, who had taken my course a year previously, was relating his not-yet-completed plans to leave traditional science for a year of work in Chinese holistic medicine, with particular emphasis on nutrition. He was clearly concerned about methods of evaluating the two careers with respect to social worth and metaphysical depth. His philosophical sophistication did not allow him an easy move into a new mind-set, yet he was clearly unhappy with a certain emotional emptiness perceived in his current activities.

Thinking about previous experiences with individuals who had entered life-styles with new and noncongruent ways of thought brought into focus the full paradox of deciding between alternatives. In order to meaningfully grasp a mode of existence, one must immerse oneself within the framework for a necessary period of time. This requires belief and discipline. By the time an individual gains enough experience to evaluate a metaphysical system of beliefs, he is asking questions within the vocabulary and thought patterns of that mode, and the answers will be contingent upon the questions. Acquiring an intimate familiarity involves sufficient commitment to compromise one's objectivity. This is truly the paradox of paradoxes.

Upon being presented with this conundrum, my young companion readily agreed, but he challenged with the thought that the same line of reasoning was equally applicable to science. This idea, while not a new one, was here presented in a particularly concrete context: an individual career choice. It is undeniable that being a scientist, or any other professional, requires a long novitiate of dedication, trust, and hard work before one can comprehend the full meaning of the material. After that training, successful candidates undoubtedly raise questions framed within the context of their special disciplines. Detailed methods of validation are not external to but lie within each branch of a particular subject.

The paradigmatic nature of science has been much in the fore since the first appearance of Thomas Kuhn's *Structure of Scientific Revolutions*. Examining groups of scientists from a sociological perspective is a first step in reducing their views of the universe from an absolute to a more relative status. The contemporary philosopher Paul Feyerabend in his polemic work *Against Method* goes much further in downgrading the special status the natural sciences may have as contrasted to any other systems of belief. He goes deep within the history of science to show the roles that subterfuge, rhetoric, and propaganda have played in the acceptance of one view over another. Feyerabend's extreme and often irritating book formed a backdrop for our attempts to devise a scheme to compare the validations of traditional Oriental medicine and biochemistry.

I felt frustration that the arguments were not clearer. We know, of course, that science represents mankind's best efforts to seek out the truths of the universe. Yet we are aware of these self-evident truths from within, and our clear and certain knowledge has no special objectivity when viewed from the outside. When faced with these difficulties, we feel as if the wily Feyerabend has set a trap, tricked us into questioning what we know to be true, and forced us to compare our chrome-plated, computerized, digitized, analyzed, and commercialized hard knowledge with the crazy, hazy, lazy mutterings of wandering mendicants. It is downright embarrassing.

These enigmas indicate the desirability of an epistemology that eliminates the need to be within a discipline before we can evaluate it. I do not know if such a philosophical technique exists, but it is part of the faith of a rationalist that such an approach can be developed. There are no objections to pluralism per se, but I cannot get over the idea that we're all in this world together and had better learn, at a minimum, to communicate about our experiences.

There is no way to predict which path this student will take. "Two roads diverged in a wood," and he alone must decide. We parted, understanding each other, with him voicing concern about the paradox of paradoxes and how it concerned his world view. In truth, I must admit that despite a firm and long-standing dedication to the scientific enterprise, I can sometimes hear the ghost of the paradox whistling in the wind outside the chambers of my mind.

St. James Infirmary Blues

Now that computers are here to stay—and they *are* here to stay—one suspects that the issue of artificial intelligence versus the human brain will be with us for the foreseeable future. A few events of recent months suggest the exploration of some odd byways of this age into which we are speedily moving.

After some years away from the San Jose–Palo Alto area of California it was impressive to see the enormous growth of population and business. If the streets weren't lined with gold, they did appear to be shadowed by silicon, water thin. Microelectronics had moved from the laboratory into the mainstream of industry and home life, and here was the center of that action. Somehow I sensed no great collective *joie de vivre* on the part of the local residents. The tempo is too fast to savor the joys of being on the forefront. The time for a corporation to move from a garage or basement operation to Wall Street—or at least an over-the-counter listing—may be as short as five years. At such a pace there is little leisure to meditate on the philosophical challenges of artificial intelligence. However, the St. James Infirmary in Mountainview (between San Jose and Palo Alto) was still serving cheap drinks and free peanuts, so a familiar oasis was available wherein to ponder some of the eternal queries. It seems characteristic of modern times that, although novel inventions become industrial very fast, there is still a long lag in incorporating scientific and technological ideas into our view of man, society, and the universe. The work of the metaphysician is still slow and painstaking.

One Chivas Regal and a number of peanuts later I was able to focus my musings on two reports that had recently crossed my desk: "The Great Electronic Mouse Race" and an article by my friend Hans Bremermann entitled "Complexity and Transcomputability." The mouse race, sponsored by the Institute of Electrical and Electronics Engineers, offers a prize of $1,000 to the builder of a self-contained mechanical mouse that can negotiate a maze in the shortest possible time. The "mouse" is allowed three passes at the maze and is timed on the final try. In the first preliminary competition a five-inch-long, five-inch-wide, seven-inch-high "mouse," powered by alkaline cells, had a time of 51.4 seconds. The secret of the robot's success was microcomputers, a good wall-sensing probe, and a properly engineered set of turning wheels.

The field of electronic rodents is in its infancy, and there is no telling where it will lead. However, since psychologists have traditionally used maze learning in mice as one measure of intelligence, mousedom is clearly threatened by technological unemployment, as well as the assaults on the psyche associated with being replaced by machines. There is something new under the sun, and the classic problem of building a better mousetrap is now being joined by the problem of building a better mouse in order for the world to beat a path to your door.

The article on complexity and transcomputability is found in *The Encyclopaedia of Ignorance*, a work whose very title set us to pondering about contradictory ideas. Bremermann looks into limitations on the information processing capacity of computers. This question has related software and hardware answers. A computer carries out sequential operations following an algorithm, or set of rules, for the solution of a problem. Many computations, when approached in this manner, require so many individual steps (of the order of 10^{100}) that the solution can be envisioned only by enormously large and very high-speed machines. Here the hardware limitation enters because of fundamental physical restrictions. The speed with which information can be transmitted from one place to another is limited by the velocity of light, according to Einstein's special theory of relativity. Thus if two components are a certain distance apart, the shortest time per operation must be larger than that distance divided by the velocity of light. The obvious answer of making computers smaller and smaller is constrained by the atomicity of matter and the random thermal motion at the atomic and molecular levels. A second limitation comes from Heisenberg's Uncertainty Principle, which requires that the faster an operation takes place, the greater the energy uncertainty, and the larger the amount of energy that must be contained in a signal to assure us that it is read without error. In summary, the rates of information processing are tied to the most profound relations of matter and energy. The amazing job of miniaturization that the microcomputer engineers are doing has its limitations deep in the laws of physics.

Since the hardware problem appears to be fundamental, the ultimate approach would seem to lie in the direction of better algorithms. A branch of mathematics known as complexity theory has been formulated to deal with this problem of computational costs in terms of the number of steps. Many algorithms are found to be transcomputable. Their implementation exceeds the ability of any computer that can be constructed by the known laws of physics. Many problems in artificial intelligence can never be solved by contemporary techniques.

When faced with this enigma many years ago, von Neumann conjectured that the central nervous system must use a kind of mathematics entirely different from any we have discovered. Such a mathematics would not be governed by present theorems on complexity but would require new software and perhaps new hardware. We must face either the epistemological consequences of Bremermann's limitations or the challenge of von Neumann to discover a totally different type of analysis. Both prospects are intellectually very exciting.

These reveries were interrupted by the waitress with another Chivas and the arrival of an old acquaintance. When I unburdened my thoughts on him, he recounted his recent embarrassment at having been beaten at chess by a computer that cost only $800. We resolved to push the problem of mice, man, and machine to its ultimate limit by designing a robot with the task of enjoying good scotch.

The device, which we named MacSilicon, is quite thrifty and requires only one tenth of a cubic centimeter of brew placed in its reaction chamber. After the sample equilibrates with its vapor, an aliquot of that vapor is injected into a gas chromatograph mass spectrometer, and the results are recorded. After the sampling, the vapor pressure is brought up to a preset value by raising the temperature, and the process is repeated. The primary data consists of a series of mass spectra, and a program converts these into chemical compositions. A large number of fine scotches are "drunk" by the machine, and the data are stored. When an unknown sample is consumed, it is compared with all of the scotches in the memory bank, and the machine registers a numerical estimate of its enjoyment based on an algorithm computing vector distances between the new scotch and others it has been taught to enjoy. The printout on scotch data tape will tell us how much the machine is enjoying the drink. The thrifty robot need take only one sip of each kind of scotch, for it gets as much pleasure in consulting its data bank as it can in downing another shot of the same stuff.

At this point the thought exercise was complete. We lifted our glasses and toasted each other, for in our heart of hearts and mouth of mouths and stomach of stomachs we knew that the machine would never experience what we were experiencing. Secretly I wanted to take a tenth of a gram of peanuts and jam it into MacSilicon's reaction vessel.

Do Bacteria Think?

In ruling last June that new genetically engineered strains of bacteria may be patented, the justices of the Supreme Court seemed to imply that the tiny organisms were not fully alive in the same sense that higher organisms are. Like paper clips or Wankel engines the bacteria were a product of "non-naturally occurring manufacture and composition of matter." What the Court did not seem to realize was that this materialistic view runs counter to recent developments in microbiology. Researchers are turning up more and more evidence that these single-cell organisms are living, sensate beings whose activities can be described with such psychologically oriented terms as stimulus, response, excitation, and adaptation. The new studies raise a profound question about the evolution of mind: If the simplest forms of life are capable of purposive activity, can they be said to engage in a form of thinking?

The issue is posed by studies of what is called bacterial taxis, the migration of these organisms toward or away from certain chemicals, oxygen, or light. While the phenomena have been reported since the late 19th century, the recent revival of interest has produced unanticipated information that has established behavioral biology of bacteria as a research discipline. Its proponents believe that their studies will yield insights into the behavior of more complex organisms.

The physical act at the basis of single-cell behavior consists of swimming from one location to another under the influence of environmental signals. Motile bacteria tend to be rod-shaped and are equipped with a number of long, helix-shaped flagella at their sides or ends. Tiny biological motors in the membranes of the cells rotate the flagella, providing the motive power to propel the rods through their liquid environs. The motors can rotate in two directions; one direction produces smooth swimming in straight lines, while the other leads to a tumbling motion. Bacteria normally alternate between swimming and tumbling. The resulting trajectory is random in direction, somewhat like the walk of a drunk who proceeds in a straight line but periodically changes direction, thus following a zigzag path with no overall direction.

The stimuli for nonrandom behavior by bacteria are environmental changes in the concentration of certain molecules called repellents or attractants. If a cell moves toward a higher density of repellent (usually a toxic substance), it tumbles frequently, thus changing direction. If it moves toward a higher density of attractant (usually food), it tumbles less frequently. This behavior fits the stimulus-response model of classical psychology, with the chemical change as the stimulus/input and the frequency of tumbling as the response/output. Such consistent movement toward food and away from poison represents a bit of evolutionary wisdom that is of clear benefit to the organism.

Part of the interest in microbial behavior reflects an effort to assess where on the evolutionary tree certain psychological phenomena originated. For a long time there has been a peripheral branch of psychobiology that has searched for signs of learning in progressively smaller and more primitive organisms. Since the great bulk of psychological research involves such higher vertebrates as mammals and birds, only a small corner of the vast literature contains studies of presumably simpler systems.

Among the more primitive invertebrates examined, the annelid worms appear to show maze learning, classical conditioning, and habituation (a decreased response to continued stimulation). Classical conditioning has also been reported in flatworms (planaria), a group whose behavior received a good deal of faddish attention a few years ago. All the invertebrates in which learning has been reported possess, at a minimum, a neural network—an interconnecting set of nerve cells with a signal-processing center.

There have been several attempts to demonstrate learning in organisms without any nervous system. Some of these experiments show habituation, but many investigators would question whether that constitutes true learning, because it could result simply from tiring on the part of the organism. Plants of the mimosa family have leaflets that close when they are touched or otherwise stimulated. After repeated stimulation the leaflets cease closing. A similar failure to respond to repeated stimulation seems common in protozoans and fungi.

Other observations, made outside the laboratory, confirm the feeling that even the tiniest creatures are very purposeful in their behavior. For example, in the beautiful pond-water movies of Roman Vishniac, the small creatures swimming about pursue prey, mate, avoid obstacles, and flee predators with remarkable repertoires of action. Although there is a considerable gap between observed behavior and thought as we know it, Vishniac's camera has unmistakably captured the sense in which these tiny creatures are aware of their surroundings.

This seeming demonstration of psychic activity as we move down through the phyla still leaves us somewhat unprepared for the idea of behavior in bacteria, which are vastly smaller and simpler than even the fungi or protozoans. Through studies of biochemistry we have come to regard these organisms as chemical factories—a point emphasized in the Supreme Court ruling mentioned earlier. In contrast to the Court's ruling, the behavioral studies tell us these bacteria are sensate beings that not only feel things in their environment but also respond to sensory inputs.

The precise way that swimming cells detect their surroundings has been studied, with the fascinating conclusion that bacteria sense changes in the space around them by measuring variations in time. Because cells move at constant speeds, the temporal variations measure what is happening in their space coordinates. The ability of a single cell to detect differences in time requires, at least, a fast chemical reaction with the attractants or repellents to reveal present conditions and a slow internal reaction to reflect the past state. The ability to sense time requires memory, another psychological term that has entered the molecular vocabulary.

Swimming bacteria are able to recognize about 30 different attractants and repellents, evolutionarily chosen from the vast array of chemicals to which the species has been exposed. Since the response to each is different, the final behavior must integrate these varied responses. This allows for a sophisticated assessment of the opportunities and dangers in the habitat and provides the cell with the best pathway through them.

One feature that emerges from this research is the remarkable similarity between the mechanism by which bacteria sense different chemicals and the analogous process by which the cells of higher animals sense and respond to the presence of neurotransmitters and hormones. This similarity reinforces the optimism of workers in the field that they are opening the way to a better understanding of behavior in high organisms.

This continuity of behavioral process from the simplest of organisms to the most complex engenders two radically different responses to the question, do bacteria think?

From the point of view of the biochemical determinist, bacteria do not think. Rather they respond to stimuli in the environment, using known chemical principles. Their reactions are integrated by a programmed signal generator that triggers a response in the rotary motors of the flagella. Following this line of reasoning, fungi do not think, protozoans do not think, and mimosa do not think. But if this is true, where does thought as a distinguishable feature arise in evolution? The most consistent materialists say that it never arises. Annelids do not think, planaria do not think, and invertebrates do not think. They respond to signals with a response/output whose usefulness is tested by evolution. This line of reasoning leads inexorably to the conclusion that Supreme Court justices also do not think but simply respond to stimuli in a manner that has passed the evolutionary filter for survival.

Mentalists—believers in the existence of mind—would argue from the continuity of behavior to the opposite conclusion. Since we are able to move step by step from Supreme Court justices, who we know can think, down the evolutionary ladder to successively simpler forms, then some sort of psychic activity must be ascribed even to the lowliest organism, bacteria.

We are left with the dilemma of having to accept one of two conclusions: Either bacteria think or Supreme Court justices do not. The only way out is to assume that at some level of organization between microbe and man, thought arose as a new phenomenon. This theory would present us with the fascinating task of finding the first species that demonstrates true psychic activity and might inaugurate an evolutionary psychology or psychological taxonomy. There is no telling where such reasoning might lead. At the least, the observation of behavior in bacteria extends the science of psychology throughout the living world.

ESP and dQ *over* T

The acronym *ESP*, standing for "extrasensory perception," has become the most widely known term from the general field of psychic phenomena. Of all the borderline effects, ESP comes closest to scientific respectability by being perceived as similar to the paradigms of normal science. However, confusion exists because ESP has two quite different meanings. One school of thought believes that information may be transmitted from one individual or object to another individual or object by means of physical signals that we have not as yet discovered. These carriers may be electromagnetic waves in some little-studied spectral range, or gravity waves, or some other type of ill-characterized energy transmission. Such a postulate has the virtue of being within the framework of contemporary physics. Other advocates of ESP believe that transmission involves methods that are totally outside the range of measurement of physical devices and are not energy dependent in the thermodynamic sense.

The first concept of ESP does not fit the ordinary meaning of the words, for if a physical signal is sent from a source with a sending device to an individual with a receiver, there is nothing extrasensory about the process. De facto it is sensory; we have only failed so far to locate and characterize the sense organ. Such phenomena may be very interesting, but nothing revolutionary is being explored that is likely to change our philosophical concepts of science. I would suggest that we use the term *USC*, "unidentified sensory communication," to describe such phenomena. This may cause some confusion with the University of Southern California, but usage doubtless will make clear which USC is intended.

The second type of ESP, or true *extra* sensory perception, is an entirely different kind of idea. If it is proved to exist, it will seriously alter our ideas about physics and biology. This kind of ESP would, as we will show, violate the second law of thermodynamics and force a basic reformulation of most of science.

To examine this discrepancy between psychic phenomena and normative science, we go back to the mid-1800s, when Rudclph Clausius and William Thomson were formulating the second law of thermodynamics, which states that spontaneous processes always tend toward a maximum of molecular disorder. The measure of this submicroscopic chaos is designated entropy, and changes in this quantity are computed from the changes in the heat, dQ, divided by the temperature, T, under appropriate conditions. Thus an increase of entropy is associated with all spontaneous processes.

Shortly after the original findings about entropy increase, the British physicist James Clerk Maxwell posed the following paradox. He imagined two boxes of the same gas of equal volumes and equal pressure with a wall standing between them. In the wall he assumed a tiny trapdoor operated by a minuscule creature. When the being saw a molecule approaching the door from the right and no molecules approaching from the left, it opened the door; otherwise the door was closed. As this process was repeated, an excess of molecules was built up in the left-hand chamber and a deficit developed in the right. Carrying this out for a sufficient time would result in all of the molecules on the left and none on the right. The hypothetical creature became known to science as Maxwell's demon. His devilishness consisted of defying the second law of thermodynamics by spontaneously changing the state of the gas from the disordered uniform distribution to the more ordered state showing a concentration difference.

The annoying demon hung around in physics for a long time, until he was exorcised by Leon Brillouin using the powerful methods of information theory that had just been developed by C. E. Shannon. Brillouin reasoned that in order to know when to open the door, the demon would have to make observations, that is, take measurements on its environment. The demon would require photons or some other physical signal to see the molecules. Such signals require the expenditure of energy. Brillouin then was able to show that this power expenditure leads to at least as much molecular disordering in the measurement as can be gained by opening and closing the door at the appropriate times. The crux of this scientific argument is that you cannot get something for nothing, even information; energy must be expended in learning about the state of the system. This reasoning, which was put in more detailed mathematics by information theorists, resolved the problem that Maxwell had set over half a century earlier.

Now assume the possibility of extrasensory perception. In the problem set by Maxwell it would correspond to a demon who would know the state of approaching molecules without any accessory disordering reactions. This would of course violate that revered second law. And as everybody who remembers that $PV = nRT$ will be quick to see, a difference in the number of molecules in the two boxes is equivalent to a difference in pressure. Such a pressure differential allows a force to be exerted on a piston and permits mechanical work to be done. After one round of such work the demon could once again engineer a pressure difference and power could be generated. Thus if true ESP exists, we can design a device to continuously convert the heat of matter into work. Among other benefits to mankind the energy crisis would immediately be solved.

By this time you are probably looking at your leg to see if some demon author is pulling it. Not so! In plain old-fashioned physics extrasensory perception, if it exists, could counter the laws of thermodynamics and usher in a golden age of energy. The Maxwell demon is not an essential part of the argument. Any knowledge of the state of a system obtained without taking a physical measurement lowers the entropy and opens up the previous possibilities. All right, supporters of ESP, the ball is in your court.

While musing over these matters, I chanced upon an article reporting on a number of secret CIA memos from the 1950s that had just been released under the Freedom of Information Act. One such document revealed that the secret agency had considered using ESP to spy on unfriendly powers. This suggests limitless possibilities, a "mind tap" without a court order and countless other invasions of privacy. In any case, this memo demonstrates how far psychic thinking has crept into the establishment mentality.

Come, come, gentlemen, I thought, are we to be a nation of "second law and disorder," or are we going to flirt with alien ideologies, alien even to the laws of physics? With ESP and the laws of thermodynamics, we can only have it one way or the other. There is no happy medium.

The Hit Parade

There are some of us best described as "number lovers": people who just cannot resist columns of data. This love of the statistical manifests itself in many normal and perverse ways. Imagine the joy of such a figurative addict when presented with a table entitled "Relative Frequencies of Occurrence of Composers' Works in Musical Performances" (originally published in *Théorie de l'information et perception esthétique*, Abraham Moles, 1958). Listed are 100 composers and the percentage of listener hours devoted to their musical creations in concerts and operas. The pulse quickens as the eye leaps over the list, avidly looking for one's favorite. There he is: Tchaikovsky, number 8, with 2.8% of listener hours (not over our stereo, where he seems to come in at about 33⅓% with the "1812 Overture" alone).

So as not to prolong the suspense, the big seven are Mozart (6.1%), Beethoven(5.9%), Bach (5.9%), Wagner (4.2%), Brahms (4.1%), Schubert (3.6%), and Handel (2.8%). Operatic composers have a tremendous advantage in this competition, since the scoring is done in listener hours and operas are by far the longest works normally performed. Mozart is number 1 because *Don Giovanni, The Marriage of Figaro*, and *The Magic*

Flute occupy 1.1% of all listener hours, a startling fact to contemplate.

After Tchaikovsky in the ratings come Verdi, Haydn, Schumann, Chopin, Liszt, Mendelssohn, Debussy, and Wolf. These 16 giants of the music world account for over 50% of all listening time of classical music. I do not know nor does the book state how these data were collected. However, since Moles later notes that Toscanini rushed through Beethoven's Ninth Symphony in 63 minutes, 25 seconds, while Furtwängler required a leisurely 73 minutes, 50 seconds, to do the job, I can only conclude that the investigator must have attended numerous concerts in order to compile this table. One gains further respect for the background research upon finding more detailed information indicating that Toscanini took a full minute and 10 seconds longer on the Scherzo, in between beating out Furtwängler by four minutes on the Allegro and five minutes on the Adagio. In any case, I hope that some contemporary is willing to update this work and provide us with an ongoing account of civilization's musical preferences. What could we learn, for example, by comparing such data for American, Soviet, German, and Italian listening audiences?

Looking at the leading composers, I am suddenly embarrassed. There is a total lack of familiarity with number 16, listed simply as Wolf. This is quickly remedied by finding a brief biography of Hugo Wolf (b. 1860, Slovenj Gradec, Yugoslavia; d. 1903, Vienna). He composed over 300 lieder, in addition to a number of symphonic poems. A music critic as well as a composer, Wolf's short and tempestuous life appears to have ended in an asylum. Make a note to purchase a recording of some of Wolf's work. Number 16 should not be a stranger to the ear.

Number 17 is familiar enough: Jean Sibelius. One of my first musical memories is of attending a concert in which my mother played violin in an orchestra performing a tone poem by the great Finnish composer. Upon reflection, this list of musical genius virtually sings with childhood melodies and memories: "Papa

Haydn's dead and gone, but his music lingers on" and "Morning is dawning, Peer Gynt is yawning, and the music was written by Grieg."

Scanning the assemblage, one notes two Strausses and three Bachs. While the Strausses are unrelated, the Bach family constitutes a true musical dynasty, with father Johann Sebastian joined by sons Johann Christian and Carl Philipp Emanuel. Each son accounts for about 0.2% of listener hours, and they rank 91 and 92 in the standings. Actually, it is most impressive that the offspring could have stood in the shadow of the great Johann Sebastian and achieved substantial musical fame in their own rights. The elder Bach was apparently a great teacher as well as a great musician. Make a note to read about Johann Sebastian and Maria Bach and how they raised their family.

And what about the 100th name on the list, the man who just made it into the charmed circle? It is Giuseppe Tartini of Padua (1692–1770), a violinist who composed over 200 sonatas and 200 concertos. He is best known for the violin sonata "The Devil's Trill." He also made a contribution to science by noting the difference tone, or third note, that can be heard when two notes are steadily played. The phenomenon of acoustical beats appears to be the discovery of this prolific musician.

In addition to the 100 leading composers, the works of 150 others are occasionally played in concerts and account for about 6% of all listener hours. In the history of classical music about 250 individuals have dominated the field—allowing for regional preferences, perhaps as many as 350. Western civilization produces great composers at the rate of about one a year on the average. Of course, this is just a statistical figure, which clearly fluctuated every time Maria Bach called the midwife. Virtuosity in composition is a rarity, and the overwhelming popularity of the small group at the very top makes this point even more emphatic.

Speculation about genius is one of those endlessly fascinating leisure hour games for admirers of creativity. For individuals not gifted with musical ability the very thought of composing a symphony is awe inspiring. My more talented friends assure me that it is equally awe inspiring for most musicians. The idea of being able to intellectually synthesize the output of 80 instruments and hear in one's mind the final result seems incomprehensible, if not downright mystical. The continued productivity of Beethoven after becoming deaf reinforces our feeling about the enigmatic character of musical composition. Symphonic music, this deep and nonverbal mode of communication, must stand as one of humanity's great triumphs.

There are endless possibilities in the analysis of the quantitation of musical success. For example, 18 of the 100 leaders have last names beginning with B, an event of very low probability, considering the distribution of first letters of European names. Then there is the analysis by country of origin; my impression is that the Germans dominate and the English are underrepresented. For the more mathematically sophisticated there are rank-frequency distributions. For the musically informed there is analysis in terms of style and type of composition. The possibilities are clearly vast. By now readers will be beating down the doors of their libraries to see this information for themselves. The final bit of good news is that there is an English translation of Moles's book published in 1966 by the University of Illinois Press. The data in question are on pages 28 and 29.

For dedicated music lovers, I must heartily endorse this kind of research. Imagine spending one's hours sitting in concert halls, gathering data with computer terminal, stopwatch, and metronome. Lincoln Center, the Sydney Opera House, perhaps Milan —it will be necessary to collect a variety of input from around the world. Make a note to apply for a grant to study the mathematics of musical preference, and be sure to include tuxedo and opera glasses under capital expenditures.

SECTION FOUR

PRISON OF SOCRATES

Some Social Issues

Prison of Socrates

It is not my usual custom to write an essay while sitting on a rock on the Acropolis; nevertheless the reasons are compelling. The architectural splendor of the Parthenon and surrounding ruins by themselves might inspire poetry rather than this hesitant prose, but I have just come from the supposed site of one of enlightenment's earliest martyrdoms, the prison of Socrates. I had experienced some difficulty in finding the exact place. There are no signs, no monuments; only ΣΩΚΡΑΤΕΣ scratched in the stone, graffiti-like, informed me that I had arrived at my destination. The path to the prison started at the foot of the Acropolis and passed near a small church that echoed with ageless liturgical chanting. Drawn to the chapel door, I was distracted from my mission for a few minutes of listening. Nearby stood a small restaurant where a kindly waiter, with whom I did not share a common language, responded to my inquiry after Socrates by escorting me to the porch and pointing out a walkway leading up a hill.

The prison cell was the simple two-room cave described by Plato. Iron grillwork closed off the entrance, and I peered inside, feeling a strange admixture of past and present. The dimly remembered words of Plato somehow emblazoned themselves on the here and now, creating a surrealistic timelessness: "Were you yourself, Phaedo, in the prison with Socrates on the day when he drank the poison?" "Yes, Echecrates, I was."

It is strange that the location of such an important happening should go unmarked. In the half hour that I was there, not a single tourist appeared to experience this shrine. Could it be that the descendants of the citizens of Athens who voted the death of Socrates do not wish to advertise the matter? That is a charming notion, but it is probably more ideal than real. Perhaps there is too much uncertainty about the authenticity of the location. I don't know, and I am fully prepared to leave the mystery unsolved, being content with the good fortune of visiting the place in tranquility, uninterrupted by the constant comings and goings that characterize such a popular attraction as the Parthenon. One of the chief reasons for including Athens on our itinerary had been to see the prison where the great philosopher had met his noble end. The motivation was a long-time attraction to Socrates' doctrine that the beginning of wisdom is the realization of how little we know. It is a necessary antidote to the tendency to regard ourselves as wise and to consider present knowledge as the final word.

Still somewhat dazed by having walked on the stones that had known the sandals of the famous philosophers of antiquity, I climb the Acropolis and look out over the modern city of Athens. In my mind's eye I try to sense the scene as it must have been in the days of Pericles, but something obtrudes on that fantasy. Hovering over the city is a dark cloud of pollutants so obscenely dirty that it blots out all visions in a reeking miasma of filth. I have seen that cloud floating over many great cities. Yet viewed from the Parthenon, it takes on a new significance, for in looking back at the past, we are tempted to look toward the future. The events of the morning are forcing me to move from the Golden Age of Greece, to modern Athens, to some dim beyond.

It is simply not true that the past is good and the present is bad, as some romanticists would maintain. The trial and death of

Socrates is in itself evidence enough that classical Greece was something less than a utopia. Yet there is an interesting contrast between one lone man sipping poison for ideological reasons and an entire city of several million slowly inhaling a lethal mixture as the price of a life-style committed to certain technological and material wants that have become needs. Even here the comparison is informative. What is done for modern Athens by pollution-causing machinery was done for their ancestors by slaves. The simple notion of good and bad begins to slip away, and we focus instead on the issue of what a modern society wants. How have we arrived at these goals? What price are we willing to pay?

We can, in this context, begin to understand what made Socrates so different and classical Greece so unique. He dared constantly to raise the questions: What is justice? What is the good life? What are the responsibilities of the state? By obstinately placing these difficulties before his fellow citizens, he kept them from falling into the rut of merely accepting a way of life for its own sake. He forced each Athenian to look inward, even if just a little, and ask why he was doing the things he was doing. Modern Athens lacks a living conscience, a public nuisance who will force the citizens to pause and question the most fundamental assumptions of their lives.

Classical Athens was what it was because most of the citizens knew Socrates personally. Present-day nations are so large that individuals know only their small circle of acquaintances, and the state is an abstraction that can be comprehended but not touched emotionally. Most of the great social writings of antiquity deal with relations of individuals at a one-on-one level of interaction. The Ten Commandments deal with parent and child, man and wife, neighbor and neighbor. These interactions, while immensely important, have been overshadowed by social patterns that relate the individual to his world in a statistical and impersonal way. We are now faced with problems where each person's contribution is minute, but the aggregate is potentially disastrous. The black cloud over Athens is such an issue. Concerning these moral dilemmas, the most cherished writings of antiquity stand mute, since the problems have only arisen with the industrial revolution and the enormous population growth of the last 200 years.

If wisdom does not provide us with answers, it does suggest that we should not cease from asking the questions. In the performance of this task our societies are lacking. We turn our attention to such pragmatic features as manufacture and transportation, but we do not inquire what we should be making or where we should be going. A highly diverse culture will no doubt offer many answers to such challenges, and arriving at conclusions will be exceedingly difficult. It is naive to suggest that a few simple Socratic queries and replies will suddenly turn us on to the right road. I do maintain, however, that if we do not constantly labor over the major points of social ethics, we will surely miss the mark, for the guidance of the past does not fully encompass the present.

The world is in need of annoying, troublesome, Socratic-like thinkers who will keep us from intellectual and spiritual slumbers brought on by lethargy, hyperstimulation, self-satisfaction, or simple discouragement over the magnitude and complexity of the challenges that have been set before us. Such philosophers are needed in education, journalism, television, movies, and every other public forum. They will trouble us and cause us sleepless nights, and I suppose that from time to time we shall imprison them or worse. But in the end they are national treasures, and if their graves or the sites of their martyrdom are unmarked, their ideas are the catalysts that enliven life and keep us from stagnation.

It is a very emotional experience sitting here on the Acropolis, and I wish not to let the event pass without finding deeper meaning in life. "Tell me, Reader, what do we mean by a good society?"

Turning Colorado into Kansas

In 1973 the state legislature of Colorado enacted a statute (34-1-305) setting forth the doctrine that no governing body shall "permit the use of an area known to contain a mineral deposit in a manner which would interfere with the present or future extraction of such deposits by an extractor." Ordinarily such action could simply be chalked up as a victory of the mining interests over the conservation interests, and the matter would be no different from thousands of other conflicts between political and economic forces. The uniqueness of this case can be seen in Special Publication 5-A of the Colorado Geological Survey, which regards "sand, gravel, and quarry aggregate resources" as mineral deposits. Once limestone, dolomite, basalt, rhyolite, quartzite, granite, and gneiss are included as minerals, all of the Colorado Rockies become potential grist for the stone-grinding mills. Given these concepts, one begins to envision the possibility of great mountains being pulverized and the plains of Kansas spreading westward toward Utah, propelled not by some geological process but driven by the hand of man. I do not by this comment wish to denigrate the topography of Kansas. In itself it has a certain two-dimensional beauty, which may have inspired such literary works as *Flatland* and echoes with the profound themes of

Euclid's *Elements*. The feeling remains, however, that the two neighboring states gain by geographical contrast, and any attempt to homogenize them would be a net esthetic loss for all. It is just too difficult for those of us who are over 30 to imagine Dorothy and Toto coming from Colorado rather than from the plains of Kansas.

The burden of the mineral legislation appears to have fallen most heavily on Jefferson County, which lies to the north and west of Denver. Located between the gravel-hungry urban areas and reaching to the foothills and front-range mountains, this county is under the most intense pressure to allow quarrying, blasting, and crushing. The city of Golden, the county seat, is the center of commercial plans for these operations. Currently one gravel producer within the city limits is removing the top half of a mountain. Another nearby project is consuming part of the Dakota Hogback, a unique geological formation.

Applications have been filed to zone several more mountains in the immediate region of Golden to allow for extraction operations. The state statute has an extremely limiting effect on the county's prerogatives. Thus the citizens of Jefferson County are losing the right to determine their own zoning, even if that zoning interferes with the most exploitative of all possible uses of the land, its removal. Local autonomy, in the form of the county's own well-formulated mineral extraction policy plan (June 1977), is coming into sharp conflict with the enactment of the state legislature.

Jefferson County (how appropriate that it bears the name of the author of the Declaration of Independence) is thus a microcosm of the global conflict between society's use of its resources and the interests of local residents whose turf is being chewed up. At first sight this looks like a clear-cut case of good guys versus bad guys, but like most great societal conflicts, the battle of Jefferson County is not that simple, for we are reminded in the previously mentioned publication 5-A, "Sand and gravel are basic to the construction of our homes, schools, hospitals, churches, shopping centers, streets and highways, airfields and bridges." We are further told of the role of these resources in sewage and water treatment, agriculture and industry. These materials clearly are the core of the building industry in rapidly growing urban areas.

Human exploitation of natural resources dates back to the first hominid who picked up a rock to throw at some other animal. Forests have been leveled for farmlands, and earth-moving operations have been an integral part of the development of human society. The present situation differs from that of the past in that our engineering prowess keeps increasing the magnitude of these operations, and we are now chipping away at the lithosphere itself: We are influencing in a major way that rocky skin that coats our planet. Indeed, a geological fault line passes through Jefferson County, and some earth scientists at the Colorado School of Mines in Golden have expressed concern that the quarrying may set off seismic disturbances of a major character.

The surface of the earth has undergone continuous change over the last four billion years. Continents drift and great ranges rise and fall. Seashell fossils found on top of tall mountains are telling evidence of the shifting of geological formations. Before the age of human engineering these changes were usually slower and life had time to adjust on an evolutionary scale, but the proposed rate of alteration allows no such biological leisure. The basis of a geological-ecological ethic cannot depend upon whether we will exploit the planet; our very existence ensures that we will. The core of our moral problem is how much is too much; it becomes a more difficult quantitative ethical dilemma and requires at the outset recognition of the emptiness of all-or-none propositions. Denver will in one way or another have its sand and gravel, for to stop crushing the rock would be to start crushing the economy. Within that framework we need to develop environmental policies for this situation and for all others.

Two major issues stand out in this type of confrontation. The first is: At what level should major habitat decisions be made? Are they the province of the local residents, who will bear the noise, the dust, the heavy road traffic, and the loss of scenic beauty, or are they the province of larger governmental units, which must balance the needs of more diverse geographical areas? Here our Jeffersonian democracy would argue for utmost local autonomy, with specific acts of the higher legislature required to override a political subdivision's decision if such override is deemed to be in the interest of the larger governmental unit. This is just the opposite of the present Colorado situation, where county governments are constrained by a blanket statewide law. It seems clear that statute 34-1-305 is in obvious violation of the principle of local self-determination.

The second issue—how much is too much, or how far is too far—is at the very core of a planetary ethic. A categorical imperative is needed to spell out a rational balance between use and abuse. If I am unable to enunciate a clear doctrine to cover these cases. I am also not free to give up seeking such a principle. The problems of nuclear waste, recombinant DNA, pollution, and destruction of the landscape all fall within the province of evaluating the lengths to which we may go in pursuing certain goals at the risk of known and unknown hazards. The lack of an axiomatic ethical framework within which to generate quick answers to such pressing problems causes frustration. Their immediacy also deprives us of the luxury of deferring an answer until after we grow wise. Given these quandaries, the wisdom of moderation stands as a beacon to maximally avoid irreversible processes. Land use policy is our covenant with our planet. It is an obligation to keep up the work of understanding responsibility within that context. If such understanding does not emerge in time from the ivory towers, it will be faced in the courts and legislatures, where great questions of this type are ultimately decided in a practical sense. In the meantime, the Rockies of Colorado and plains of Kansas both are treasures, and every effort should be expended to preserve them. If we are to "Let freedom ring from every mountainside," then it is unambiguously clear that we will require mountains.

Welcome, Class of 1984

The faculty wishes to extend a warm welcome to entering freshmen, the class of 1984. We have long awaited your coming with some trepidation, because George Orwell has made you a group apart by setting his futuristic novel in the year of your graduation. He did not, by his writing, wish to set the mark of Cain upon you; he was more humane than that. Rather he tried to present a warning to us about what might befall the world were we to relax our continual opposition to all varieties of doublethink and doublespeak.

The novel, which now falls heavily on your shoulders, was one of a negative utopian genre that appeared from the 1920s to the 1940s. The best known are *Brave New World* by Aldous Huxley and *Nineteen Eighty-Four* by George Orwell. Although they failed to predict in detail the present state of Western culture, they did offer chilling warnings about the endpoints of certains trends. Huxley and Orwell both were unhappy prophets (cheerful prophets are rare), who ethically rejected the values they forecast. While their visions were vastly different, the two shared an intuition about one very significant root cause of the worlds they described.

Orwell reveals his analysis of that cause through the writings of the fictional character Emmanuel Goldstein, archenemy of the state. In Goldstein's words, "Ever since the end of the nineteenth century, the problem of what to do with surplus of consumption goods has been latent in industrial society." Here we see the radical and oracular nature of the novelist's thought. While conventional economists were clamoring for more productivity, Orwell anticipated that the industrial and technological revolutions would lead to a glut of material goods. Going beyond economics, he saw this excess creating exciting human potentials of social equality and leisure time. In the Orwellian nightmare, however, world leaders perceived these two possibilities as threats to their own privileged positions and responded by waging continuous wars among the three superpowers in order to consume the productive capacity, keep everyone busy, and maintain a hierarchical social order with themselves on top.

Huxley was much less specific about economics, but he, too, forecast a state of great productivity and vast amounts of free time for the most intelligent citizens, the alphas and betas. This leisure time was filled with an ongoing orgy of controlled sex, drugs, and entertainment, all designed to reduce reflective thought to an absolute minimum. The ethic of the highly ordered *Brave New World* is playing one's assigned role without philosophical questions or emotional commitments. Social equality is prevented by biologically creating castes that resemble the divisions of an insect society more than those we regard as human. (Have the sociobiologists secretly been reading Huxley?) Both societies were designed to suppress, by carrot or stick, the free exercise of thought by capable individuals.

Why did these futurists see leisure time and equality as so destabilizing to society? Free time permits free thought, which among bright young people may characteristically take a revolutionary turn, threatening the values of a society. The status quo is always endangered by thinking people with the leisure to pursue that activity. Equality, the second threat, somehow violates some hierarchical instincts that we have inherited from our primate ancestors. While we accept equality as a moral human goal, it runs counter to a lot of monkey business that we carry around in our heads. A society with true equality would force this issue into open conflict.

The good news I have for you is that the grimmest excesses envisioned by the two futurists have been avoided. Although a few years ago the West, the Communist bloc, and the Third World were beginning to look like the fictional Oceania, Eurasia, and Eastasia, we have stepped back and avoided the worst features of ongoing war. The living standard of the West is far different from, far better than, the dismal picture painted in *Nineteen Eighty-Four*. Although some college campuses have occasionally resembled the drug and orgy culture of *Brave New World*, this is only one view of a multifaceted institution. Reflective thought remains an important part of our educational process.

Although we have avoided the worst, the related problems concerning productive capacity and leisure time, as seen by the novelists, are still with us. And unless we learn to deal with them, they remain possible threats to economic stability and social tranquility. Human adaptability has not failed, however. We have, in fact, almost without foresight, lucked into a much more humane solution than portrayed in the anti-utopian novels.

The modern world has taken the ancient Chinese institution of bureaucracy and encouraged it to grow to such huge proportions that it is capable of consuming any amount of excess productivity. If production gets too high, we have the ability to decrease the number of workers and increase the middle-management level. This growth of bureaucracy has paralleled the growth of industrial society and has been established simultaneously in free enterprise and in socialist societies. Curiously, it knows no creed, no race, and no economic system and is independent of place and ideology. Such a ubiquitous institution must be doing something right.

It is the fashion in all developed countries to deny overproductivity and to condemn bureaucracy and wage campaigns to reduce it. But these are obviously mythical and symbolic gestures, for the institution grows and flourishes. Government commissions on paperwork continue to issue massive reports. To understand why this happens, we must go back to Goldstein's analysis that it is necessary not only to destroy excess productivity but to do it in a psychologically acceptable way. War accomplishes this by providing an emotionally rooted reason for destruction. Bureaucracy provides individuals with a raison d'être for decreasing the output. Each bureaucrat is supremely sure he is performing a socially useful function, indeed a necessary function, so he is able to block productivity with the psychological assurance that he is a productive individual. And if equality and leisure time are threats, as Goldstein assumes, then the functionary is correct: He is aiding society by the limitations he imposes. Indeed, if you agree, I suggest you take a bureaucrat out to lunch.

Thus, students, your 1984 will not be stabilized by Orwell's war or Huxley's ongoing orgy; rather, it will be cybernetically controlled by paper—huge masses of multiple copies, computer printouts, and undreamed-of reams. These will be discussed at meeting conferences and in endless rounds of organizational communications. The millions of people carrying on this activity will all have titles assuring them of the importance of their efforts. The entire process will regulate the productivity of society and preserve the hierarchy. Leisure time will be controlled by busyness, lest it be occupied with such destabilizing social activities as thought.

You, too, can participate in this world and, because of your education, rise to an important post in the hierarchy. With the skills that we will teach you, a nobly titled job will be available in 1984 and you will be able to join your countless peers who will be taking their places in this structure.

But how about those atavists among you, throwbacks to an earlier age, who do not wish to prepare yourselves for the conventional 1984; What can we as an educational institution offer? Fortunately you will find professors who will teach you about philosophy—that agonizing search for the real that began on the streets of Athens and has wound its way tortuously through history on a road strewn with enigmas, fallacies, and visions. You will find teachers of literature who will introduce you to other vistas, joy, existential pain, martyrdom, and, on occasion, startling insights. You can interact with scientists who are trying to probe the mysteries of the universe with the same passion as their humanistic counterparts. But what will you do when you graduate in 1984?

Because you will require sufficient leisure to wrestle with your thoughts, you will be different from other graduates. You may have to operate outside the mainstream institutions. If you choose—and have the fortitude—you may struggle with the aim to be creative and productive within the Establishment.

You may help solve the overproductivity problem by creating esthetic, intellectual, and pleasurable products that can be absorbed by a society without limit. You may labor to convert leisure time into a satisfying, rewarding, and ultimately stabilizing social institution. You are free to educate people to equality as a goal, not a threat to their positions in the pecking order. We are not forever bound to be the kind of people who inhabit a *Brave New World* or a *Nineteen Eighty-Four*.

You will probably not have the easy life of your classmates working within the bureaucracy. There are many products of industrial society you will learn to do without. However, there are two things I assure you that you will have: my endless admiration and respect and the opportunity to create a far, far better world than Orwell and Huxley envisioned in their negative futurism.

The Infernal Combustion Engine

Christian Huygens was, as far as I know, a thoroughly exemplary scientist. He discovered the rings of Saturn, invented the pendulum clock, and was a leading exponent of the wave theory of light. Then around 1680 he made some kind of Faustian deal and began playing around with inventing the internal combustion engine. His version didn't work because he used gunpowder and had difficulty in maintaining continuous cycles. But work or not, the secret was out, and by 1860 Etienne Lenoir developed a thoroughly practical version of Huygens's brainchild. The world has never been the same since, for with the coming of the small gasoline engine, the precious gift of silence has all but disappeared from the earth.

Like many others, I had grown so used to the phenomenon that a pair of earplugs had become an extended part of my anatomy when I sat down to write or think. Then a curious manuscript crossed my desk one day. It was written by a man who had gone to spend a year living in solitude in a little handcrafted hut on the shore of Squantz Pond in western Connecticut. With the author's permission and his name withheld according to his wishes, part of that manuscript follows:

After I had moved into my hand-built house late in autumn, I resolved to spend a portion of my leisure each day listening to the sounds of man and nature and attuning my ear to the symphony of life as it unfolded through the seasons. I awakened the next morning to the sharp, brisk roar of two chain saws as some neighboring woodsmen prepared for the coming winter cold. I saw a flock of migrating birds but could not hear them, for the honk of the wild goose, while cacophonous enough, is no match for an unmuffled one-and-a-half-horse engine. I grew to look forward to the chain saws in the morning, for they reminded me of the cycle of man's labor.

On the seventh day the woodsmen rested from their labors, and I awoke to an eerie silence. It was so quiet I could hear the cars on the highway a full quarter mile away. My thoughts on the automobile were interrupted by the sounds of my neighbor's leaf blower, which went on for the next four hours. When he stopped, I could still detect the low moan of distant leaf blowers and an occasional leaf mulcher. Most sounds are seasonal, and I listen carefully to the leaf machines, for I will not hear them again until the earth has orbited the sun one more time.

In *Triad*, Adelaide Crapsey has written:

> *There be*
> *Three silent things:*
> *The falling snow . . . the hour*
> *Before the dawn . . . the mouth of one*
> *Just dead*

It is winter, the last leaf has flown before the vortex of the rotary blade, and I sit listening for the silence of the snow. It comes at night and my first notice of the floating crystals is the deep bellowing of the highway department snowplows rushing along, clearing the roads and dumping sand. The blade of the plow along the road has a peculiar music all its own as the heavy steel bump, bumps along the concrete.

With the coming of dawn the white blanket brings a new beauty to the world. The trees hang heavy with snow, and along the shore I can see the snowmobiles as they rev up to dart in and out among the trees, following the forest trails. They remain for the entire day, indeed as long as snow is on the ground, and hardly a daylight moment passes when I am not aware of the small vehicles whizzing happy people through the world of nature.

A few weeks later the wind changes direction, and I find myself in the landing pattern of the Danbury Airport. A whole series of new melodies descending from on high alerts me to look up and watch the planes fly by. I am becoming quite an expert at identifying them from their calls. Occasionally a helicopter flies up from Stratford, and the special dissonance of its tone alerts me to rush out of the cabin lest I miss this rara avis.

With the melting snow many familiar voices disappear, and I can once again hear the cars along the highway. Spring is approaching. "For lo, the winter is past, the rain is over and gone. The flowers appear on the earth; the time of the singing of birds is come, and the voice of the turtle is heard in our land" (Song of Solomon). I know the birds are singing, for I once again see them, but the sounds I hear are the Hell's Angels roaring their Harley-Davidsons down the highway. As the ground dries, even they are drowned out by Yamaha trail bikes skittering over the paths just recently abandoned by the snowmobiles.

Spring mobilizes my neighbors, and the weekends resonate with hedge trimmers, lawn mowers, pruning saws, and edgers, all singing the song of the oscillating piston as it goes up and down in the cylinder in a sort of endless sexual frenzy.

Summer now comes, the sun crosses its highest parallel, and laughing children leave their school to frolic near my humble cabin. I watch them and marvel that such young tykes can handle such massive, thundering speedboats. Now the sounds come closer as the boats pulling water-skiers whiz by my home. Some days there are so many boats that I envision myself as living inside a kettledrum of the New York Philharmonic during a performance of Rimsky-Korsakov.

The deep-throated boats are joined by a higher-pitched sound as the young folks play with their gas-powered model airplanes. I marvel at how so much sound can be generated on such a small tank of gas.

Now autumn comes again, and I have spent a year in isolation at my cabin. I have seen and smelled and heard the comings and goings of the seasons. I have not thought too much. My brain seemed to oscillate with my eardrums, and I could never decide whether to gear up the left side or the right side of my cerebrum.

I am longing for contact with people again. Today I will leave here, and tonight I will sit in the stands surrounded by cheering humanity as I watch and listen to the West Haven stock car races.

Legacy

One of the most painful aspects of being a teacher is to catch a student cheating. The more optimistic among us try to reform the offender and set him on a firmer path. The following letter is true, and the story is unfinished.

Dear ———.

Now that it is obvious to both of us that your term paper was pure plagiarism, we simply cannot let the matter pass without discussing it further. My personal displeasure and disappointment need be of no special concern to you, since the social issues loom larger by far than the personal ones. We both exist inside a scientific subculture whose none-too-permanent foundations are built on an ideological commitment to be honest with nature and with each other.

It is strange and, as you shall see, most ironic that you chose to cheat in the manner you did. For in seeking a source to copy, you managed to come upon a little-recognized paper that I personally regard as one of the most important works of our time in the particular area of science under consideration. Had you but credited the original author, I would have been delighted by your exposition, for choosing this article as a source shows a profound understanding of the subject matter on your part. Many contemporary scientists do not grasp the meaning of this study, which you comprehend very well. Part of my sorrow in contemplating your moral lapse is that it is conjoined with such intellectual facility. The combination shakes my sense of cosmic balance.

The irony goes much deeper. I was personally acquainted with the author of the paper in question, and a few years ago I was deeply saddened by news of his suicide. In part this untimely end resulted from a depression induced by his ideas being ignored by his peers. Thus your recognition of the man is not without its salutary aspects. Let me therefore tell you a bit about the scientist who produced the original work.

As a student this individual began to question certain foundations of the subject matter he dealt with. He realized that the explanations of his professors were often glib and facile and covered over certain real uncertainties in the root assumptions. (I have noted in you that same questioning attitude.) As he pursued his scientific career he spent more and more time struggling with the unresolved problems and trying to clarify them. He was much less successful than his peers, who simply accepted the structure of the subject as given and went on to the applications. He was not a quick thinker but deep and excruciatingly honest in his pursuit of truth. His genius was perhaps not up to the task he set for himself. In the context of Greek tragedy this would have been his fatal flaw.

After years of very hard effort he did indeed come up with some extremely important insights. They were written down and published in the paper you chose to honor by dishonoring. It is at best a difficult piece to read. Each idea was wrung out of such struggles that the author was unable to present the material without including the scars of those battles. I spent about two months, on and off, thinking about this work and then wrote to the author to congratulate him. Only a few people around the world appreciated what had been accomplished.

In his home institution our scientist was either ignored or regarded as an eccentric, wasting his time on "weltschmerz" rather than buckling down to the latest experimental approaches. People responded to his results with one of two comments: "It's wrong" or "It's obvious; I knew it all along." Some even responded with both comments.

In the ensuing years I corresponded with him in relative ignorance of the trauma he was undergoing. He was, however, deeply hurt. The failure to gain recognition after such exhausting effort oppressed him and in the end depressed him. He continued to work, but the spark was gone.

About three months before his premature demise I visited him. He lived far away from me and was pleased to have a distant visitor. His insight was wanted on a scientific problem I was wrestling with, and I was not disappointed. His suggestions initiated a fruitful approach to the theoretical quandary. He talked a bit about his frustrations and his difficulty in getting down to work. He even mentioned certain suicidal tendencies. I tried to reassure him in the only way I could, by reaffirming my belief in the great importance of his contribution. I said it as loudly and clearly as I was able. He spoke somewhat plaintively of the successful scientific dynasties, such as the Darwins and the Huxleys.

When I returned home, there was a letter with some extending remarks about our scientific conversation. My reply to these remained unanswered, and two months later word arrived that he had taken poison. I do not know what other factors entered into his decision. Every suicide is something of an awesome mystery. But it seems apparent that in a certain sense he was the victim of his own scientific honesty and integrity. There is no happy ending to this story. He was in his mid-30s when he died. In his desk lay an unfinished book-length manuscript detailing his views.

You can now perhaps begin to sense my feelings when I received your version of this man's work represented as your own. These ideas were not originally grasped quickly out of thin air for a term paper; they were truly won by sweat and tears and in the end by blood. You had no way of knowing how truly grand your larceny was, but that which you stole was very precious.

I cannot recommend for you the hair-shirt honesty and dedication of this man; it is too difficult a path. But if you place your actions in apposition to his, surely you will see that these events can have a purpose if they place before you an example to ennoble your life as a scientist. I have personally been enriched by having known a very, very honest seeker of knowledge, and by your actions you have come to experience his influence. Coincidence is a fickle agent, and the more mystical among us would feel that the fates were acting herein to redirect your life. Let us just say that you are very fortunate to have this model set before you. This time you have sold out cheaply; our exemplar paid the ultimate price for his integrity. You are young and the future is before you. The choice is yours.

Sincerely,
Your teacher

Facts and Artifacts

It was a quiet evening meal in the dining room of the Hall of Graduate Studies. Four students seated at a small circular table were engaging in the talk of the day. At the meal's conclusion a microbiologist among them stood up and announced, "I've got to go up to the lab and get a culture started." Two social scientists in the foursome looked puzzled at first and then burst into laughter. Their confusion was caused by the multiple meanings of a word. The speaker was talking about transferring a drop of fluid from a test tube of growing bacteria to a similar tube of a sterile medium. But the anthropologist and sociologist conjured up images of tribal people taking off in canoes, traveling to a distant island, and establishing a new way of life.

The participants in this little vignette were being treated to an example par excellence of the varying and highly specialized use of language in the biological and social sciences. The situation recalls C. P. Snow's "two cultures" theory, which concerns the lack of communication between the natural sciences and the humanities. The reference to Snow comes quickly to mind because the events centered on the very word *culture*, which is central to his viewpoint. The social scientists continued to conjure up images of the many ways of starting cultures, and dinner ended in good-humored repartee.

In addition to *culture*, other terms cut across the natural and social sciences, but none in so telling a way as the strange word *artifact*. The dictionary definition, "a thing made by art, an artificial product," hardly explains the varied passions that the word evokes among archeologists and biologists.

For those scholars interested in the study of human societies and civilizations, an artifact may be more precisely defined as an object showing human workmanship or modification. Anthropologists often focus on tools, weapons, jewelry, and totems of a tribal group in efforts to relate behavior and technology. Their studies rely heavily on the products of the human hand. Archeclogists digging into the past must depend almost entirely on artifacts as they attempt to reconstruct early cultures from what remains. Those objects, which were produced by the workers and artisans of the early civilizations being studied, are the primary factual materials for archeological research. As such, these artifacts are welcome, and the word has a joyous connotation for social scientists. To discover an artifact is to move one's field ahead, which is a major aspiration for any scholar.

For biologists, on the other hand, *artifact* is a dirty word, an epithet, even an insult. The implications of the term and some clue as to the reactions may be seen in the entry from *Encyclopaedia Britannica*: "Artifacts, in biology, the appearance of unnatural structures in chemically and physically prepared animal and plant tissues as a result of the acts of preparation." The modern usage in molecular biology goes beyond this narrow context and defines *artifact* as any spurious observation caused by preparative procedures.

A historical example illuminates the situation in the biological sciences. In 1883 Camillo Golgi (later Nobel laureate in physiology) used chemical stains that he had devised to demonstrate certain otherwise invisible structures in nerve cells. They became known as the Golgi complex or Golgi apparatus. Similar structures were later demonstrated when the same staining procedures were applied to other cells. For more than three quarters of a century arguments ensued as to whether the Golgi apparatus was an artifact of staining or a real cellular organelle. The advent of electron microscopy finally resolved the issue to most scientists' satisfaction, and the Golgi apparatus is now included in the catalog of cellular hardware.

What is so curious about the definitions of biological artifacts given above is that the crucial phrases "unnatural structures" and "spurious observations" are almost completely subjective. Precise usage would dictate that almost every observation made in the laboratory is of an artifact in the dictionary sense of that word. We slice tissues, stain them, grind them, extract them, and act on them in a thousand other ways to generate products that are very far from the original state of the living material. Our final observations are related to structures in the living organism in a way that depends on a complex line of reasoning, which makes many assumptions about how the living material has changed in each step of the processing. What gets evaluated by scientists is not usually the quality of the observation per se but our confidence in the theoretical assumptions that connect the living material and the laboratory data. Thus I judge what I am seeing through my microscope to be a fact and what my colleague is seeing to be an artifact if his conclusions disagree with mine. The old adage that seeing is believing has to be subject to a lot of amendments.

Artifact as used by the biologist is a euphemism for "you don't understand what you're doing." It is hardly strange that the term has become an exceedingly unpleasant one, sometimes uttered in venomous hisses at meetings of learned societies. Since the number and complexity of intermediate steps between living cells and publishable data keeps increasing, it is likely that the use of this epithet will continue to grow. Science at the forefront is a complicated business, and the resolution of issues often requires a continuing confrontation between contending viewpoints. It is difficult for disputants who hold strong opinions to be dispassionate and uninvolved. To keep normal discourse going in such charged situations, highly symbolic words like *artifact* serve as socially accepted devices to insult an opponent.

Other disciplines have also adopted *artifact* with varying emotional overtones. Dermatologists use the word to denote a self-induced wound, and we have references to cases of *dermatitis artifacta*, as in "Don't pick on your pimples or you'll get *dermatitis artifacta*." Electronics engineers seem to have adopted the spirit of the cell biologists' meaning. They refer to spurious signals as artifacts whether or not these signals are generated by humans. Thus they strive to develop amplifiers free from the annoying artifacts caused by self-generated random voltages. In this usage they have moved totally away from the human aspect to something caused by the disordered motion of electrons, a universal feature of the material world. I think the engineers have gone too far in extracting the annoying aspect of artifacts without considering the root cause. They have, in fact, been careless with language.

The same word with the same root meaning has taken on very different connotations and emotional responses in different areas of scholarship. This discussion of words has, then, turned up some sociological features. We have moved from word usage by certain scientists to a consideration of why professionals use words as they do. We end up by looking at the patterns of culture of people who start cultures, study them, and send messages about them. All of this takes us back to the Hall of Graduate Studies and the marvelous chance for cross-fertilization between the sciences and the humanities that occurs whenever experts in various fields of study dine together. This recalls the case of two physicists spending an entire meal discussing the measurement of what they call a magnetic moment. A student of French literature at the next table, having heard bits of the conversation, finally leaned over and said, "Magnetic moment. My, what a wonderful name for a perfume." That's probably not exactly what C. P. Snow had in mind, but we have to start somewhere.

SECTION FIVE

DEALING FROM A FULL DECK

Unusual Individuals and Their Work

Dealing from a Full Deck

While witnessing history in the making, one is rarely aware of the full significance of the events taking place. Preoccupation with the nuts and bolts of everyday existence usually keeps us from pausing and reflecting on the long-term aspects. So it was, many years ago, when I was busily arranging slides to be shown at a conference on information theory in biology to be held at the National Institutes of Health. The ringing of the phone was followed by a voice on the other end booming forth in distinctively Russian accent: "This is Gamow." George Gamow, Professor of Physics at George Washington University, was a familiar name because of his work in astrophysics, his popular science writings, and a persistent rumor that he was working on the genetic coding problem in biology. He continued his conversation: "Can I come to give a talk on Friday?" There was no question about it; if Gamow wanted to talk, Gamow could talk. I realized that my own time on the program would be cut in half to create a slot. But never mind, everyone would be eager to hear what the famed scientist had to say about the relationship of polypeptides and nucleic acids.

There was insufficient lead time to get his talk listed on the printed program, so his introduction as final speaker of the session was a happy surprise for many of the attendees. The professor, a large man, was an imposing figure as he mounted the podium and began the presentation. Contained in his subject matter was the first detailed theory of the specification of amino acids by the nucleotides of DNA. Basing his argument on probability theory and analyses of amino acid composition of proteins and base ratios of DNA, Gamow postulated a relationship between the four-letter alphabet of DNA bases and the 20-or-so-letter alphabet of amino acids in proteins. On a theoretical level molecular biology was being moved forward by a very large step, although few were aware of it at the time.

What curiously remains in memory from that day is the fact that Professor Gamow could not, or would not, remember the names of the purine and pyrimidine bases: adenine, thymine, guanine, and cytosine. Instead he presented his theory in terms of spades, hearts, diamonds, and clubs. It may have been pure showmanship or a strong desire to clear the essentially mathematical argument of biochemical detail, but for whatever reason playing cards were used throughout. The analogy was useful. Just as there are four nucleotide bases, there are four suits; corresponding to purine bases and pyrimidine bases there are red and black cards. The speaker went all the way and brought in polymer models made of single-card monomers. His mode of presentation made the talk memorable, and the subject matter made it timely. In historical perspective we see that it was an important lecture.

The theory was of course wrong in detail. It assumed a direct DNA protein code, without consideration of the subsequently established role of RNA. It did, however, force people to focus on an abstract triplet theory of coding and so stimulated a flood of thinking and experimenting, which eventually culminated in our present understanding. Unknown to us at the time, Gamow had communicated his theory to Crick and Watson, who with a number of other researchers had begun to seek refutations and tests of the basic ideas.

Professor Gamow was certainly one of the great generalists and popularizers of our time. He is a member of a generation of mathematical physicists who seem to have moved easily from one field to another with something significant to say in each discipline. Leo Szilard went from nuclear physics to molecular biology, with asides in politics and science-fiction writing. John von Neumann made notable contributions in mathematics, quantum mechanics, computers, self-replicating automata, and government administration. Werner Heisenberg is still read by both philosophers and physicists. This is but a sampling of a group devoted to an intellectual comprehension of all of nature. These scholars grew up and matured in the excitement of the emerging quantum mechanics, and they had great confidence in the ultimate rational solution to problems. Their influence on our intellectual world is very deep.

Another characteristic of these scientists was a sense of humor. Although never doubting the seriousness and importance of what they were doing, they pursued their goals with an élan that frequently broke through into a smile. Gamow was ebulliently and boisterously humorous, the very antithesis of one public image of the savant on the forefront of research. Some of those following today seem so terribly somber that one wishes these contemporaries gave more evidence of enjoying their work. Science is truly exciting and uplifting, yet the heavy hand of bureaucratic and administrative procedures is making research seem like grim business. The very success of the enterprise renders it less enjoyable to pursue, and it will be a great loss if the most catalytic young people are driven away from the field. We must counter this tendency by being able to laugh at ourselves and our "benefactors." Gamow's use of playing cards was also telling us that science should be fun.

Sometime after the original presentation of triplet coding, our hero, George Gamow, was talking before an American Physical Society meeting in Baltimore. A large audience had gathered because the coding problem was of central interest to a number of society members turned molecular biologists. The original talk had become more polished, the theory had advanced, but still the hearts, spades, clubs, and diamonds were the core of the presentation. A malfunctioning public-address system challenged the full resonance of the speaker's voice. At the conclusion there was time for discussion, and one questioner rose to inquire, "Tell me, Professor Gamow, why a *triplet* code?" The speaker stroked his chin, and after an extended silence he replied, "As my grand-uncle who was archbishop of Odessa would have said, God loves a trinity."

It seems odd that so many concepts in modern biology have been developed by physicists. There is a tradition, going back at least to Immanuel Kant, of individuals trained in the exact sciences making notable contributions in other fields. A striking example is Franz Boas, who went from physics to leadership in anthropology. However, these are isolated cases, whereas in the immediate post–World War II era the migration from physics to biology was quite a substantial social phenomenon. The roots of the movement are seen in part in the 1930s, when Niels Bohr raised questions about the applicability of the new quantum mechanics to biology. In 1944 Erwin Schrödinger, viewing biology as a "naive physicist," emphasized the notion of a genetic-code script and the covalently bonded character of the genes. Dr. Gamow, who had worked with Bohr in Copenhagen, was acquainted with these views and emerged for a short time as a leader among the DNA-transformed physicists. This generation of scientists seems to have exhibited a few common features. Their training retained much of the natural-philosophy aspect that characterized the 19th century. When World War II ended, many of them wanted to change the direction of their research into more humanistically oriented endeavors.

While several of these researchers showed a wide range of interests, few, if any, seem to have exhibited the breadth of talent of George Gamow. His subjects moved from radioactivity to the "Big Bang" theory of the universe, from the origin of chemical elements to genetic coding. In addition, he wrote popular scientific books in which he did the illustrations. While his accomplishments are most substantial in many fields, for some of us who attended the NIH Symposium on Information he will best be remembered for portraying the coding nucleotides as playing cards. A certain tinge of jealousy enters as one sits memorizing complex biochemical formulas. One thing is certain: In the game of life, here was a man who truly dealt from a full deck.

As the World Turns

Museums are more than just repositories of the past; they also serve a scholarly function in the pursuit of new knowledge. In addition, the exhibits may evoke responses at an emotional level that stem from a deep curiosity about our world. As an undergraduate I occasionally wandered through Yale University's Peabody Museum in an effort to gain some insight into past, present, and future. Standing before the magnificent "Age of Dinosaurs" mural by Zalinger, one was tempted to focus on the evolutionary time scale, so vividly presented, and place current concerns in proper global perspective.

At the entrance to the museum hung a Foucault pendulum, which, by its continuing swings, demonstrated the rotation of the earth on its axis. This device was modeled after one first set up by J. B. L. Foucault in the basement of his house in Paris. The original was a five-kilogram weight suspended from a two-meter steel wire and free to move in any direction. Set to swinging along the earth's meridian, it made a mark on the floor indicating its plane of motion relative to the planetary surface. While the earth turned, the direction of the pendulum's plane of motion fixed, remained fixed, owing to the conservation of angular momentum. As a result, the line of motion changed relative to the floor, which was firmly attached to the earth and therefore varying in its angular position. At 2 A.M. on January 8, 1851, the first successful demonstration of the earth's rotation by this means was carried out. The inventor went on to erect a much larger version of his planetary clock in the Pantheon in Paris. This remarkably simple apparatus provided a precise and beautifully visual demonstration of one aspect of our planet's motion.

Each time I passed the display, I stopped and thought just a little about the celestial clock before me. No matter how ominous the human condition might seem, before my eyes was irrefutable evidence that the earth was still turning on its axis, and in some cosmic sense the world was in order. It engendered a sort of Emersonian tranquility about the universe. I must confess that a suspended weight makes a strange security blanket, but that is apparently the price we must pay for intellectualizing our existence. Over the years frequent visits to the giant metronome always elicited an upbeat reaction.

Then the unthinkable happened. One day about 20 years ago I entered the museum and found the Foucault pendulum gone. In its place stood a booth for selling tickets of admission to the museum. In a state of disappointment and disbelief I sat down next to a triceratops skeleton and plotted strategy. The logical response seemed to be the organization of a mass movement to reverse this unconscionable decision. A faculty-student protest needed to be organized. In a fantasy world I began to envision hundreds of young idealists and their professors blocking the door of the building and shouting, "Give us back our pendulum." Other long-haired radicals carried signs pronouncing, "Foucault Lives." My mind's eye saw one student in blue jeans with a t-shirt bearing the motto, "Free Angular Momentum." But, alas, at that time we were in the midst of an unpopular war, and a generation gap was tearing the nation apart. One could not, for selfish reasons, organize a civil disobedience with potentially riotous results. Plans for a student movement were abandoned as being too irresponsible.

Leaving the triceratops, I wandered over to the remains of some

Eocene mammals and tried to gather my thoughts in order to formulate a fitting and nonthreatening type of protest. A letter to the editor of *The New York Times* seemed proper. In thought it commenced: "Dear Sirs: I am writing to you about a matter of some gravity." With these words I faced the realization that I would not be able to convince the public at large of the seriousness of my pleading. Those who had not experienced the deep emotional impact of standing in silence before the seemingly unending oscillations would not quite understand.

I wandered around from exhibit hall to exhibit hall looking for a cheering thought or at least a possible path of action. There came to mind the approach of John Milton, who celebrated each serious event in his life with a poem destined to impress future generations with the righteousness of his cause. Now standing before a woolly mammoth, I withdrew a notebook from my pocket and began, in imitative Miltonic fashion, to pen the words:

Jean-Bernard-Leon Foucault
Your pendulum swings to and fro.

A blankness came o'er my mind while I was pleading with the muse for more words to commit to posterity the feelings of that moment. But she forsook me, and I was left staring at the hairy beast in front of me with the thought that he, too, was extinct.

This story of the disappearing pendulum might have remained, under ordinary circumstances, forever untold. Except for catharsis there would have been little point in reviving those bleak memories and reliving the loss of the cherished friend. My affection for Foucault has increased since I learned that, as well as being a physicist, he wrote a general science column for the *Journal des Débats*. That in itself, however, wouldn't have justified reopening this case were it not for the fact that I recently received word that the pendulum was not consigned to the flames of the smelter but has been preserved in storage and even now could be resurrected if only the university could free the funds from higher-priority items.

This news has set me to thinking: What, other than faculty salaries and parking, could be of higher priority to a great university than this symbol of the pulse of the universe? Such a device, focusing on the motion of the earth itself, should have meaning for most contemporary scholarly disciplines, both in the sciences and the humanities. Perhaps we should inscribe on a plaque and place before the machine the words of Goethe, "Time is governed by the oscillations of a pendulum, the moral and scientific worlds by oscillations between ideas and experiences."

The motion of the pendulum has at least one further lesson for those of us given to concern with academic and intellectual freedom. It is a reminder of the struggles of the founder of modern physics, Galileo Galilei. At his trial before the Inquisitor General this student of the pendulum was forced to recant his views on the motion of the earth. As he left the court, the aging scholar was reported to have whispered softly, "E *pur si muove*" ("But it does move").

Future students should be reminded, "One generation passeth away, and another generation cometh: but the earth abideth forever" (Ecclesiastes 1:4). Give them back their Foucault pendulum!

Gossamer Birds, Flights of the Imagination

*Cape Gris-Nez, France, June 12—A flying machine pedaled by an American sitting on a bicycle seat flew over the English Channel today in the first crossing of the channel by a man-powered plane. Bryan Allen, a 26-year-old cyclist, powered the Gossamer Albatross 22 miles from the English coast in just under three hours to win a $205,000 prize. The craft was designed by Paul MacCready, an engineer from Pasadena, California [*The New York Times, *June 13, 1979].*

Rarely does a letter from one's class secretary evoke visible enthusiasm. While the pleas for support are indeed justified, they always have an ironic ring to the faculty members who are being asked to contribute in order to increase their own meager salaries. However, the latest communication was an exception, announcing a class dinner and illustrated talk by fellow graduate Paul B. MacCready Jr. I had been trying to ascertain for some time if the quiet, serious young scholar I remembered sitting in physics class was the same man whose reputation had soared on wings of Mylar and Styrofoam. The invitation left no doubt; the announced speaker was going to tell us about the Gossamer Albatross and its predecessor, the Gossamer Condor.

The stories of these man-powered planes winning two Kremer prizes—the first to the Condor for negotiating a set course and the second to the Albatross for flying across the English Channel—are well known, but it is different to hear such triumphs retold in the motivating individual's own voice. And our classmate obliged with a very fine illustrated account of the building, testing, and flying of these machines that represent the culmination of 3,000 years of effort to stay aloft in a man-powered device. It was one of those rare evenings that was both edifying and inspiring—and the food wasn't bad either.

Standing outside in the cool March night air and looking up at the clouded New Haven sky, I had a chance to digest my dinner as well as to chew over the many ideas that were triggered by the engineer's soft-spoken words. Thoughts naturally drifted back to the fabled Daedalus and Icarus, who fashioned wings of wax to fly to Sicily and escape King Minos of Crete. Icarus approached too close to the sun, which melted his wings, and fell to his death in the first reported matériel failure in the history of aviation. Our cherished myth can now be proved false because waxen wings are far too heavy and unmanageable, according to aerodynamic calculations of the necessary dimensions. The Gossamer Albatross had a wingspan of 96 feet and weighed 70 pounds. The achievement of man-powered flight awaited the development of Mylar, carbon filament tubing, and steel wire of very high tensile strength. These substances represent triumphs of human ingenuity in the fields of chemistry and materials science. Some ideas are premature and, regardless of their imaginative thrust, must await the development of an appropriate technology. That is a sobering thought, even for uncompromising optimists.

Thinking of MacCready's accomplishments does lead one to a deeper appreciation of birds, bees, and bats. Evolution has managed to generate flying animals in at least three separate and independent ways in widely diverse taxa. In each case sophisticated physical principles of aeronautics have emerged to achieve flight. One thinks of frigate birds, who spend their entire lives in the air except when nesting; bumblebees, whose performance was once calculated to be physically impossible; and bats, those mammals who are our closest flying relatives. These flying machines are remarkably made of proteins, fats, and carbohydrates. Biological structures have been most plastic in adapting to the requirements of survival, and flight is one of the more spectacular examples of that principle. The ease with which various animals move through the air made man-powered flight seem deceptively easy for several millennia. However, the absence of flight muscles is a major disadvantage if one wants to fly.

The chief take-home lesson for the class of '47 was that this project was the work of individuals, not of governments, corporations, or large organizational units (in later stages the project received financial support, without attached strings, from the Du Pont Corporation). The important elements were imagination, dedication, and a certain élan vital that motivated an exceptional group of enthusiasts. One has the distinct impression that MacCready, his family, and his friends were having fun during this venture. Their mode of operation allowed the greatest flexibility and the least commitment to preconceived ideas. In the Kremer award contests they beat out several very technically proficient teams that started by using the airplane as a prototype. The Gossamer Condor designers took as their initial idea a powered hang glider. The investment in winning first prize was $60,000 plus uncounted volunteer hours and immeasurable spirit. It seems clear that a larger organization could not have accomplished the goal of man-powered flight without expending 100 times the hard cash that generated the winning entry. There is still room for individuals with "impossible dreams," and the Gossamer Condor looked as awkward as most artists' conceptions of Don Quixote's windmill. The difference, of course, is that, unlike Cervantes's hero, MacCready met with unqualified success. I am sure that there were times, however, when the whole project looked quixotic to the designer and his coworkers.

The final inspirational point came in the last few minutes of he evening's program, when MacCready, unwilling simply to rest on his laurels, talked about present programs and future plans. The primary problem faced in the design of the flying machines has been getting maximum effect from minimum power expenditure. The human body as a one-third-horsepower motor weighing 130 pounds was clearly the weakest link in achieving the desired end. The task of optimizing travel within the limits of available energy is common to all transportation systems, and the creators of the Gossamer birds have turned to this general area of engineering. The same construction principles and aerodynamic design have resulted in a two-man bicycle, which has raced over 55 miles per hour and thus has won an honorary ticket from the California Highway Patrol. While this vehicle is far from practical for normal road use, it does indicate one path toward innovative developments.

A major challenge for future societies is the limited availability of fuel. A principal reason for the prodigious expenditure of fossil carbon is that we ride around in vehicles generally rated between 60 and 350 horsepower. The technology being developed by MacCready and his co-workers is raising questions about how we can vastly reduce that power requirement and still get sufficient performance out of motorized vehicles. The potential rewards are a reduction of fuel consumption from 50% to 90% or more.

It is not possible to know in advance whether the techniques being worked out by the winners of the Kremer prize will solve the problems of minimizing fuel expenditures. What is clear is that, given the imaginative thrust that these individuals display, almost any technological problem can be reduced to a form where some sort of solution is available. With all the talk about the insolubility of problems that is being broadcast about lately, an evening with MacCready is the perfect antidote. The Gossamer birds are tangible, if flimsy and awkward, reminders of the elegance of imagination.

Dateline: Dubrovnik

Meetings on modern science held in classical settings tend to emphasize the constantly changing ways of conceptualizing the universe. When the walls of Dubrovnik, Croatia, were being erected around the 13th century, ideas on the origin of life were dominated by the traditional views of the church. Great developments in art and literature were taking place that eventually gained this Adriatic city its title: South Slavonic Athens. When young Rudjer Boscovich roamed these streets in the early 1700s, new thoughts were in the air. Among these innovations was the mechanics of Newton, and Boscovich, who later became a leading astronomer and mathematician, was among the first continental scholars to accept the gravitational theory of Sir Isaac. The Serbian-Italian savant carried out his science as a member of the Jesuit order.

And now Dubrovnik is the site of an international conference on self-organization in biology. Novel concepts of our age are being debated just outside the medieval ramparts of the old city. Though the contrast is great, a second theme, that of continuity of culture, also emerges, stimulated in part by the very architectural surroundings.

Although ideas about the organization and origin of life embodied a severe contradiction in the 1200s, the citizens of this city obviously had other things on their minds. While biblical narrative made it clear that all living species were brought forth by God during the fifth and sixth days of creation, adherents of this view simultaneously accepted the daily origins of flies, mice, and similar forms from filth and detritus. The experimental approach to the problem of spontaneous generation did not begin until the walls of Dubrovnik had stood for 400 years. Across the Adriatic and inland from Venice, the physician and poet of Pisa, Francesco Redi, turned his attention to biogenesis. By covering flasks of meat with gauze and keeping out insects, he observed the absence of maggots and concluded that these worms developed from tiny eggs deposited by flies.

The struggles between the adherents and opponents of spontaneous generation begun by Redi waxed and waned for the next 200 years. While the physics of Newton and Boscovich was moving ahead by leaps and bounds, biology was progressing at a slower pace. When it finally became clear to all that small animals and insects required parents, the ground shifted to a debate about the generation of microscopic life. The final intellectual battle was fought over the origin of bacteria, the smallest of all free-living forms. Other more violent battles fought at the same time resulted in Dubrovnik's being subjugated by Napoleon I in 1808. Another Frenchman, Louis Pasteur, eventually designed such beautiful and convincing experiments that the scientific world almost universally came to believe that all life was derived from preexisting life. His influence was so great that for almost 40 years questions about the primordial biogenesis were effectively shelved.

Those years did, however, see an enormous development of knowledge about the age of the earth and the geophysical and geochemical history of the planet. Paralleling the nineteenth-century experiments on spontaneous generation, post-Newtonian developments were occurring in physics, posing special problems for biology. The second law of thermodynamics indicated that spontaneous processes led to increasing disorder in natural systems. Since the transition from nonliving to living implied a very large ordering at the molecular level, the origin of life appeared to run counter to a strongly supported law of physics. To the north of Dubrovnik, in Vienna, physicist Ludwig Boltzmann found the answer to the paradox. He stressed that the tendency toward disorder was a property of systems approaching equilibrium, while the living world was kept far from equilibrium by the flow of radiation to the leaves of green plants. It has taken a hundred years for Boltzmann's views to be fully accepted, but the scientific community is now cognizant that the flow of energy from a source to a sink organizes the system intermediate between those two reservoirs. The organizational principle of biology could therefore be sought deep within the structure of physics. And indeed, the Dubrovnik meeting was wrestling with just that issue.

I had come to this conference by way of the Josef Stefan Institute in Ljubljana. Stefan, honored by his countrymen in the naming of the laboratories, had been Boltzmann's teacher and a central figure in developing the laws of radiation used by his student in explaining the relation of biology to physics. Thus, I had the sense of having come full circle in searching for the physical foundations of biology.

With the advances following Boltzmann's work, there was a return to the origin-of-life problem. The 1920s witnessed an emergence from the lull following Pasteur's work. J. B. S. Haldane in Britain and A. I. Oparin in the Soviet Union realized independently that organic molecules, which might be precursors to life, could be synthesized under less oxidizing conditions than now exist. Their insights were tested in the now famous 1953 experiments of S. L. Miller and H. C. Urey on the spark-driven production of amino acids from methane, ammonia, and water. The ability to make biologically interesting molecules under conditions that might have existed on the early planet provided powerful support for further research on biogenesis.

Almost 30 years have passed since these intriguing experiments, and a gaggle of research grants have been funded for exploring the nature of energy-driven organic synthesis. In Dubrovnik we are far from the laboratory, trying to take an overview, to achieve a synthesis of ideas. There is a sense on the part of many that the thrust of the last three decades has run its course and new directions must be sought. Thomas Kuhn tells us that at such junctures scientists tend to get philosophical. It is therefore not surprising to hear words like *epistemology, reductionism, Hegel, teleology, hierarchy*, and *dualism*. These words are being uttered by developmental biologists, physicists, biochemists, mathematicians, engineers, chemists, and zoologists. An attempt is being made to communicate across the gulf that separates the individual disciplines. It is difficult work, and the success or failure of our efforts will not be measured in what happens here but in what occurs in our laboratories, offices, and computing centers when we return home.

After a particularly exhausting day we walk to the old city to sip a slivovitz and organize our thoughts. The spirit of Boscovich still lingers in our minds, perhaps as an inspiration. (Physicist John Poynting identified Dubrovnik's native son as being "among the boldest minds humanity has produced.") His *Theory of Natural Philosophy Reduced to a Single Law of Action Existing in Nature* was in many ways a hundred years ahead of its time. Some ideas in that book did not come to fruition until the theory of relativity. I find it somewhat disturbing that most Western European and American scientists know so little of this man's work. There is a kind of parochialism that shields us from some great minds of the past. As an American, I was therefore pleased to learn that in 1760 Boscovich and Benjamin Franklin met in London. One can whimsically speculate about the conversation that took place between them.

Tomorrow we return to our deliberations. Geniuses like Boscovich and Franklin are few and far between. While waiting for the next one to come along, we ordinary mortals will have to muddle through. It is, one must admit, a bit frustrating. The Yugoslavian slivovitz is, however, excellent at dulling the pain. *Zdravlje!*

On Making a Point

In a brief life of thirty-one years artist Georges Pierre Seurat lived quietly yet intensely. A contemporary of the romantic Gauguin, the passionate Van Gogh, and the debauched Toulouse-Lautrec, Seurat's existence seems pale in comparison. Rather than painting in a spontaneous or flamboyant manner, he carefully studied books on optics and art theory. He delved into sophisticated scientific works on color by such writers as Helmholtz and Rood and quickly became an intellectual leader of the Neo-Impressionists and the creator of a small number of masterpieces within that discipline.

The technical basis of his paintings, known as pointillism or divisionism, consisted of covering the canvas with small painted dots of vibrant colors that gave an intense chromatic effect. The dots were of uniform size related to the overall dimensions of the painting. His works were formalized and highly disciplined, and because of the concentration on detailed structure, they required a very long time to paint.

Some fifty-seven years after Seurat had dotted his last dot and laid down his brush, Claude E. Shannon published his celebrated article "The Mathematical Theory of Communication" in the

Bell System Technical Journal. The concept of information developed in this article has found application in diverse fields and, in a curious way, relates to art as expounded in the constructs of the Neo-Impressionist theoreticians. Shannon's theory defines a mathematical measure of the information contained in a linear array of symbols or messages, such as a series of dots and dashes, a computer printout, or a page of *Mayonnaise and the Origin of Life*. Information theory does not, as the name seems to imply, deal with the meaning of a message but with the probability of its occurrence under various conditions. A Seurat painting can be transformed into a linear message by numbering each of the dots sequentially and stating the color of each. Thus, the message might begin: carmine, magenta, magenta, magenta, pink. . . . The task of the artist in this scheme is twofold: designing and establishing the message, and executing or painting the dots. This is somewhat the way Seurat actually operated, as is evinced by the many preliminary sketches for each painting. The formal scheme enables us to discuss pointillism in terms of informational description. Each painting may be conceptualized by mapping it onto a paint-by-number set.

We can even compute the numerical value of the information measure of a major Seurat masterpiece, and this turns out to be about three and a third million bits (*bi*nary digi*ts*). In other terms, this means the artist must choose his color coding out of $10^{1,000,000}$ possibilities. This is an immense number, larger than the number of atoms in the universe—or in a billion universes, for that matter. Its very hugeness preserves art from being trivialized by technology. This is a theme that keeps recurring in relating humanistic enterprises to engineering concepts.

Information analysis is equally appropriate to paintings other than those of the Neo-Impressionists. Any picture may be divided into an array of elements and reconstructed from those components. If the elements are smaller than the resolving power of the human eye, each spot can be monochromatic and uniform, and we shall be unable to distinguish the reconstructed version from the original. Something close to this procedure is used in color television transmission, a technical field in which information theory has been of great theoretical and practical value.

Immediately after Shannon's ideas were made public, there was a burst of applications of information theory to a number of areas, including art. Books appeared attempting to apply this formalism to esthetics. Some 30 years later, information theory has taken a place of major importance in the armamentarium of communications engineers, but talk of art and information has dwindled. What became apparent after the first enthusiasm was the realization that whereas the theory can tell much about encoding a painting and transmitting it in other symbols, it clearly does not have very much to say about esthetic values.

In order for a meaningful relation between art and information theory to be developed, theorems would be required about how to generate or evaluate art. Conceptual schemes or computer programs would be needed to classify the $10^{1,000,000}$ possible paintings into categories: masterpieces, good art, poor art, junk, and noise. We could then generate masterpieces by computer graphics and retire our artists into a kind of technological unemployment. These theorems would be esthetic laws, and information theory would only be an instrument for translating the principles onto a two-dimensional grid.

The reductio ad absurdum of the previous paragraph indicates how far we are from a mathematical theory of esthetics. It might be argued by some that the very idea of such a theory is itself absurd. Attempts at formalizing art fail for two reasons, one humanistic and the other technological. Within the field of criticism there is no agreement as to a set of principles; no postulates have emerged that are generally acceptable to a broad community of experts. Thus, we do not know, in advance of looking, how to classify possible pictures that come from assigning colors to dots. The technological problems stem from the fact that $10^{1,000,000}$ is such a large number that handling all possibilities exceeds the capacity of current computers as well as that of any future computers that we can now conceive of building. Artists do not operate by systematically seeking one solution from all possible solutions. The human brain functions in the esthetic domain by processes other than those incorporated in present-day "artificial intelligence" devices.

Returning to Seurat (which is easy because I have before me a beautiful color reproduction of "A Sunday Afternoon on the Island of La Grande Jatte"), one feels that his masterpieces are somewhat independent of his theory of representation. This impression is reinforced by viewing black-and-white sketches that clearly show his talent in media in which colored dots play no part (*Seurat, Paintings and Drawings*, ed. D. S. Rich, University of Chicago Press, 1958).

Seurat was many years ahead of his time in seeing the formalism of color-dot representation. He was also an artistic genius. Genius is a mystery that we have not penetrated, and it doesn't help to pretend that we have by casually transposing scientific concepts. There are a lot of imponderables around, and while we may try to solve them, we also have to learn to live with them. We are enriched by the mysteries of Gauguin's romance, Van Gogh's passions, Toulouse-Lautrec's debauchery, and, yes, Seurat's intense theoretical formalism. Having had my say on that matter, it is with a quiet sense of satisfaction that I go back to working on my Jackson Pollock jigsaw puzzle.

Before Him Lay a Sea of Lies

The many orbital trips and safe landings of the space shuttle Columbia have revived interest in NASA's program and once again encouraged examination of the question of why we commit out national resources to the exploration of space. Before getting to this issue, I suppose that to be honest I should lay my cards on the table. I've been a space enthusiast since the comic strips and radio first brought news of Flash Gordon and Buck Rogers. At a tender age I became a lieutenant in the Buck Rogers Space Cadets. Actually, I was qualified to become a captain but lacked the proper number of cereal box tops. The rule in our house was that the top became available only after the container was empty. I couldn't force myself to eat any more of the required variety of hot mush (its name is withheld on the advice of my attorney), and my career in the Space Cadets came to an abrupt end. It's still difficult to face that particular dish, although my interest in space retains its original freshness.

Presumably, Columbia was named after Christopher Columbus to gain the full rhetorical thrust of relating the exploration of space to the world's most famous voyager. Parenthetically, I might note my irritation at our failure to refer to Cristoforo Colombo and other famous individuals by their proper names. There is something condescending about Anglicizing or Latinizing proper names rather than enjoying the full sound of the original languages. Be that as it may, thinking about the achievements of Colombo led me to read more about the intrepid Genoese explorer, whose life is recorded in great detail in *Admiral of the Ocean Sea* by Samuel Eliot Morison. The prologue to that fine biography provides a justification for my enthusiasm for space travel.

Morison begins his book with the words, "At the end of the year 1492 most men in Western Europe felt exceedingly gloomy about the future. . . . The general feeling was one of profound disillusion, cynical pessimism and black despair." He details the apocalyptic mood in a long, gloomy prediction of the future quoted from the *Nuremberg Chronicle* of July 1493. Then the mood changes, and Colombo's biographer notes: "Yet, even as the chroniclers of Nuremberg were correcting their proofs, . . . a Spanish caravel named *Niña* scudded before a winter gale into Lisbon, with news of a discovery that was to give old Europe another chance. In a few years we find the mental picture completely changed . . . . The human spirit is renewed. The change is complete and astounding."

Those few lines by Morison provide a clue to one of the principal reasons for exploring space. It is a rationale seldom heard before appropriations committees and infrequently sounded by politicians. The reason for investigating the unknown is to counter the spiritual claustrophobia that comes from a world that is spatially or intellectually closed. Mankind seems to respond collectively to overcoming challenges. At a far less global and political level, when the peak of Mount Everest was climbed, we each were enriched a tiny bit even though the triumph belonged to two men.

The counterargument, which, I admit, is substantial, states that in view of the enormous problems to solve on this planet, there is an insensitivity in committing resources to probing space. It must also be realized that much of the motivation for NASA's funding follows from military applications of earth satellites. But the negatives seem more than counterbalanced by devoting some small fraction of our effort to the kind of activity that can change the mood and spirit of the world. A society that thinks of reaching for the stars is going to be better equipped to solve its bread-and-butter problems than a society that introvertedly focuses on its darker aspects.

Reading about Colombo turned out to be rewarding in its own right. It produced several surprises, one of which found me totally unprepared. My view of the great man goes back to Joaquin Miller's poem "Columbus," which was read to me by my mother when I was four years old. Under these circumstances I acquired a somewhat idealized view of the admiral, one that avoided such matters as his avariciousness, his inordinate pride, his bastard son, and a few other of those all too, too human facts about the great man. I focused only on the inspirational "Sail on! sail on! sail on! and on!" My first surprise in reading Morison's biography concerned the number of inaccuracies in Miller's poem. For example: : "For lo! the very stars are gone" (since they were sailing at a fixed latitude, the sky stayed the same from the Canary Islands on); "My men grow ghastly wan, and weak" (the expedition was provisioned for a year, and the crossing from the Canaries took from September 6 to October 12); "These very winds forget their way" (most of the trip was due west with steady winds); "For God from these dread seas is gone" (the crew prayed three time a day and sang "Salve Regina" at night); "This mad sea shows his teeth to-night" (most of the trip was in good weather with calm seas). We have another example of why growing up is so difficult. Joaquin Miller let me down; nevertheless, I shall always be grateful for these lines:

He gained a world; he gave that world
Its grandest lesson: "On! sail on!"

The greatest surprise, and the most difficult one to deal with, was not Miller's lack of scholarship but Colombo's back of honesty. The charge is such a severe one that I quote directly from the *Journal of Cristóbal Colón* (Colombo's name in Spanish):

> Sunday, 9th of September. This day the Admiral made 19 leagues and he arranged to reckon less than the number run, because if the voyage were of long duration, the people would not be so terrified and disheartened.
>
> Monday, 10th of September. In this day and night he made 60 leagues . . . but he only counted 48 leagues, that the people might not be alarmed if the voyage should be long.
>
> Tuesday, 11th of September. . . . In the night they made nearly 20 leagues, but only counted 16, for the reason already given.

There we have it. The great man was keeping two sets of books: one for himself to tell where he really was and a second for the crew so they would not be alarmed to know how far from land they had come. (For all I know, he kept another set for the Spanish equivalent of the IRS.) Colombo had a goal and would let nothing interfere with that goal, including the feelings of his crew. That is often the way with innovators; they plunge ahead, being a law unto themselves. So I suspect some of the enthusiasts at NASA have at times had to be less than absolutely candid in order to get the congressional and public support to move ahead.

We are thus drifting perilously close to that great debate over means and ends. I must confess that after many years of agonizing over that knotty problem, I can only conclude that it is going to take a far wiser writer than I to resolve all the issues involved. In the meantime, I cheer on the space effort, for we surely need the inspiration, as did the gloomy chroniclers of Nuremberg.

Life was simpler when daily activities consisted of collecting box tops and reciting "Sail on! sail on!" But it is awesome and exciting to be probing space and ultimately finding more about our past, our future, and our cosmic connections. By a 60 to 40 vote, I'll take today over yesterday.

Warm-Blooded Fish?

A truly great book like *Moby Dick* stands the test of time because it conveys such diverse messages to its readers. Yet I venture that among the many enthusiasts few have read it as a biology text, in spite of the zoological detail that spouts forth from almost every page. On about the fifth or sixth reading, when I had become saturated with the salty sea tale and had tired of wrestling with the allegory, it seemed like a good idea to examine Melville's treatment of flora and fauna in a more directed fashion.

Chapter 32, entitled "Cetology," is in all regards a taxonomic work, an attempt at the science of classification of whales, "a ponderous task, no ordinary letter sorter in the Post-office is equal to it." Systematic biology, the work of careful and thoughtful men and women, has perhaps generated more disputes and hostile interactions than any other branch of science. Indeed, the rather proper *International Code of Zoological Nomenclature* warns: "Intemperate language should not be used in the discussion of Zoological Nomenclature, which should be debated in a courteous and friendly manner." Thus, Melville begins in a great classical taxonomic tradition by picking a fight, not with an ordinary letter sorter, but with the leviathan of systematists, Carl von Linné (Linnaeus).

The point of contention that exercises Melville is his predecessor's separation of "the whales from the fish." In a rather belligerent outburst the novelist invokes Jonah and two Nantucket messmates to arrive at the conclusion that "a whale is a spouting fish with a horizontal tail." Published in 1851, the taxonomy of *Moby Dick* does not recognize the importance of lineage, which was implied in von Linné's careful attention to physiological details, such as warm-bloodedness, breathing through lungs, and suckling the young from mammary glands.

Continuing to poke fun at the biological establishment, Melville proceeds to invent taxa—book, chapter, and species—to replace the conventional order, family, genus, and species. In his major divisions he uses a single criterion, that of size, a procedure fraught with the gravest difficulties. With literary finesse he calls his books: "I. The FOLIO WHALE; II. The OCTAVO WHALE; III. The DUODECIMO WHALE." These categories roughly correspond to large, medium, and small animals.

In this simplistic and somewhat anti-intellectual approach (a strong term to apply to a literary giant) the author of *Moby Dick* rejected the major dichotomy that had dominated whale classification since before his time. The study of the Cetacea has traditionally divided this order of marine mammals into two great suborders: the Odontoceti, or toothed whales, like Orca or sperms, and the Mysticeti, or baleen (whalebone) whales, like grays or humpbacks. Thus, the major division is based on anatomy and feeding mechanism rather than size, which has come to be regarded as deceptive. Melville discusses the biologists' approach in Chapters 74 and 75 (for those of us who get that far). Their titles are, respectively, "The Sperm Whale's Head-Contrasted View" and "The Right Whale's Head-Contrasted View." Although he was aware of this important difference from personal knowledge, the author of "Cetology" withheld it from his system, even though he noted, "The Fin-Back is sometimes included with the right whales among a theoretic species denominated *Whalebone whales* . . . with baleen." His refusal to consider baleen taxonomically (*Venetian blinds* is one of the terms he uses) is later justified by placing it among "things whose peculiarities are indiscriminately dispensed among all sort of whales."

Even though evolutionary concepts did not enter into the cetology section, a later chapter (104) on the fossil whale allows us to conclude that the author knew of the kinds of evidence that would lead to the Darwinian synthesis. In a rather personal and anti-establishment fashion Melville states his credentials as a geologist by recounting his jobs as a digger of ditches, canals, wells, wine vaults, etc. He then goes on to show a surprising erudition about fossil whales and tertiary formations, including his awareness that they are more recent than the relics of the age of reptiles. He reports on Cuvier's writing on European finds and tells of Owen's classification of Zeuglodon, a huge fossil whale skeleton found on an Alabama plantation in 1842.

The chapter entitled "Squid" establishes Melville's credentials as an invertebrate zoologist. It deals with the sighting of a great white creature, which was pursued by the Pequod's boats. When they approached, the beast disappeared. It was the giant squid known to sailors as kraken and to biologists as Architeuthis. Usually found only in the abyss, this creature provides food for the great sperm whales. Squid arms 30 feet in length have been disgorged by their predators. In the years following 1851 a number of specimens of Architeuthis were discovered and studied by marine biologists. These investigations confirmed Melville's description of this creature and its place in the food chain of the large, toothed whales. Whatever else we may know about "Ishmael," we cannot question his powers of observation or the encyclopedic nature of his background reading of everything connected with his great literary project.

These comments on the biology of *Moby Dick* are but a brief sample to indicate the magnitude and interest of this project. To paraphrase our author, they are but the draft of a draft. Large sections exist on albinism, whale anatomy, feeding frenzies of sharks, whale oil, ambergris, skeletal structure, ethology, and an array of related topics. Doubtless, years of scholarship could go into a study of Melville the biologist.

In the end, it becomes clear that Melville did not like taxonomy or formal academic biology. He commented on departments of natural history becoming "so repellently intricate." Nevertheless, he felt it necessary to place his great white whale within a system, albeit a very incomplete one. We are treated to a novelist devoting entire chapters to highly specialized and quite technical branches of science. It is part of the power of *Moby Dick* that the author did not hold back from anything that would enrich or deepen his saga. And each category of classification he introduced was enlivened by comments of a philosophical nature. Thus, when he discusses "killer," which I take to be Orca, he concludes: "Exception might be taken to the name bestowed upon this whale on the ground of its indistinctness. For we are all killers, on land and sea; Bonapartes and Sharks included." We might argue with Melville the taxonomist, but fortunately the novelist takes hold and wins us over.

How E. coli *Got Its Name*

The rise of *E. coli* from obscurity to the role of superstar of modern science must be one of the great success stories of all time. The status of this bacterium now shakes Wall Street, reverberates in the boardrooms of America's great corporations, and furrows the brows of Supreme Court justices. *E. coli* lies at the center of all recombinant DNA research and is thus at the focus of genetic engineering, a technology that is destined to alter the very foundations of our society. The current status of *E. coli* is well known, but what of its past, the humble beginnings, the early years of this Horatio Alger story? What about *Escherichia coli* even before it had gotten its name?

Our story goes back to 1885 in Munich, where a young pediatrician, Theodor Escherich, held clinical assistantships at the Children's Polyclinic and Huner's Children's Hospital. Dr. Escherich was trained in the great era of bacteriology that followed the momentous discoveries of Robert Koch and Louis Pasteur. In 1884 he had gone to Naples as a scientific assistant in the study of a cholera epidemic. Rather than confining himself to narrow clinical interests, this physician also carried out research in Otto Bollinger's bacteriology laboratory and Carl von Voight's chemistry laboratory. The pediatrician took a special interest in the intestinal flora of children as a possible clue to epidemics of diarrhea.

In 1885 Theodor Escherich was at the height of his working career. His studies of that year culminated in an original research paper in the August 15 edition of *Fortschritte der Medizin*. The contribution was entitled, "Die Darmbacterien des Neugeborenen und Säuglings" ("The Intestinal Bacteria of the Newborn and Infants"). On page 518 of Volume 3 of the *Fortschritte* we read for the first time of *Bacterium coli commune*, the original name of the present E. *coli*. Today's wonder bug entered human history in the messy diaper of a Munich infant, a truly modest start for the most widely chronicled organism in modern biology.

Dr. Escherich went on to become the leading bacteriologist in the field of pediatrics. He developed into an authority on infant nutrition and was a strong advocate of breast-feeding. He was the sole European pediatrician to address the Congress of Arts and Sciences at the St. Louis World's Fair in 1904. His work continued until 1911, when he succumbed to a cerebral hemorrhage.

Leaving the discoverer, we return to his discovery. *Bacterium coli commune* immediately became an object of considerable research interest. Sufficient information existed in 1903 so that Escherich, along with his associate M. Pfaundler, was able to write a long chapter in the *Handbook of Pathogenic Microorganisms*. In the early days a number of names were applied to this organism. In 1889 it was referred to as *Bacillus escherichia*. By 1895 the strain was variously called *Bacillus coli* or *Escherichia coli*. The matter was not finally resolved until 1919, when the genus Escherichia was firmly established in the third edition of the *Manual of Tropical Medicine*. The 34-year-old species finally had a permanent name.

Science is not such a cut-and-dried affair that it is uninfluenced by fads and fashions. Retrospectively, there is no reason to assume that *Aerobacter aerogenes* or *Bacillus subtilis* or one of very many other species could not have become the main actor in the unfolding drama. There is a certain generality in the microbial world such that most phenomena can be exhibited across a wide range of species. The choice of a single paradigm microbe tells us as much about biologists as it does about biology. It is necessary only to question the choice of this one particular bacterium among the large number of possibilities.

The appellation *coli* informs us that the organism so named is the inhabitant of the colon or lower part of the large intestine. Being one of the hardiest and most prolific of the inhabitants, it is a primary component of feces and therefore one of the most ubiquitous bacteria on the face of the earth. Because of this association with feces, the presence of *Escherichia coli* has been used since the early days of bacteriology to determine the sewage contamination of water, particularly drinking water. Since the microbial quality of drinking water has been so central to public health pursuits, testing water for the presence of this fecal bacterium was a skill known to every trained bacteriologist. For medical reasons *E. coli* was a widely studied and well-known organism.

Even before the name was firmly established, events were occurring that were destined to propel *E. coli* on the path of fame and fortune. Frederick Twort, in 1915, and Felix d'Herelle, in 1917, independently discovered the bacteriophages—viruses that attach to bacteria, enter the cells, reproduce, and ultimately destroy the hosts. Among the species found to be sensitive to bacteriophages was the ever present *E. coli*. When molecular biology underwent its grand growth period following World War II, there in the center ring was *E. coli* along with its numerous viruses.

During the late 1940s there was a resumption of fundamental research, much of which had been delayed for six or more years by the activities of Adolf Hitler. This was a period characterized by the number of individuals, trained in the physical sciences, shifting their attention to biology. These investigators wanted research materials that were simple to work with and well described. Being impatient, they also wanted a fast-growing bacterium. *Escherichia coli* neatly fit all these requirements and was chosen by many. This activity had positive feedback. Each fact that was discovered made the species a more likely subject for the next experiment. Papers on *E. coli* had a wide audience, and there are few researchers who do not appreciate a sizable public interested in their results. Thus, by inherent qualities, character, and luck, *Escherichia coli* became the superstar of the microbes.

And so it happened that Dr. Theodor Escherich has been immortalized, his name or the abbreviation *E*, appearing thousands of times each month in the scientific literature. Yet I venture that few among today's scientists have ever heard of the Austrian pediatrician or his efforts to help the children of the world. He was an idealist whose vigorous career was devoted to improving the conditions under which infants and children might grow and mature. One hopes that in the battle for the big buck that characterizes modern *E. coli* technology, the generous spirit of Escherich might survive along with the organism that bears his name.

SECTION SIX

BIOCHEMISTRY IS BEAUTIFUL

Science and Esthetics

Biochemistry Is Beautiful

The forthcoming interview by a reporter from a national news magazine was an exciting prospect for me as a young scientist. The interchange, however, went badly. Every time I mentioned adenosine triphosphate the lady balked, and we had to go back to square one. The resulting article was garbled and gave a confused account of the research in progress. The lesson I learned was "stay away from big words and be careful of what you say to reporters." The troubling issue persisting over the years is that one of the great intellectual triumphs of all time is written in tongue-twisting polysyllabic words such as nicotinamide-adenine-dinucleotide-phosphate. How is one to tell the story of this important achievement in biochemistry to a general public unfamiliar with such language? One of the most significant advances in understanding the nature of life remains unknown to most people because it is inseparable from very long words that intimidate the uninitiated and keep them from insights that hold a wide range of unexplored implications.

The intellectual accomplishment that we here praise is not the work of a single individual, nor was it put together in a blinding flash of insight. Rather, it is the product of many researchers, working at their laboratory benches over a period of more than a hundred years. Because the grand structure came about so slowly and in such small steps, few biochemists have shown interest in extolling its magnificence. Their reticence may also come from the fact that the *Chart of Intermediary Metabolism* remains an unfinished edifice, like the great cathedral of Cologne, which was left with a crane still standing on one of its towers for many years as a symbol of the tasks for future generations. Instead of being the object of poetic rapture, the network of cellular reactions is groaningly memorized by biochemistry students. The enterprise for scientists usually centers on trees, or even branches, with complete disregard of a glorious forest.

But enter almost any biochemistry or molecular biology laboratory, and you are sure to find posted on the walls or doors four printed sheets bearing a connected graph of all the major biochemical reactions that living organisms carry out in their cells. The chart is a great synthesis, a set of empirical generalizations summing up numerous experiments by generations of workers. It can be compared to other great achievements of the human intellect, such as the periodic table of the elements or the Linnaean system of classifying species. While life scientists may be silent about the deeper significance of the metabolic chart, it stands in what amounts to a place of honor in almost every research center.

To envision intermediary metabolism in proper perspective, we start with the time-honored view of biology as unity within diversity. Variety and heterogeneity are clearly evident in the array of plants and animals that greet the eye whenever we take the time to look. Estimates of the number of extant species range from two million to 10 million, and these are found in every habitat from deep oceanic trenches to the tops of high mountains. Diversity is one of those indisputable facts of life. Unity begins to emerge when we penetrate beneath the surface to the intracellular machinery and processes used by species to grow and reproduce. This examination at the microscopic level reveals common features of cell and organelle structure.

As we continue down the size range from organisms to molecules, we come to intermediary biochemistry, a collection of hundreds or thousands of enzymatic reactions by which a cell shapes matter and energy into forms appropriate for its own purposes. Here we sense the full impact of unity; the single *Chart of Intermediary Metabolism* applies with equal validity to all the millions of species that inhabit the planet. The core set of biochemical reactions of any organism from a bacterium to a great blue whale is found on the four-page chart lying on the table before me. No organism employs the full chart, but each species uses some substantial part of the reactions depicted. What may have been alarmingly complex to undergraduates studying for an examination becomes remarkably unifying and simple when we realize that it encompasses all of the diverse flora and fauna coexisting in the biosphere.

When a large body of knowledge is reducible to a compact system, scientists are tempted to look for a deeper law underlying the ordering. In the periodic table of the elements, for example, the reasons for its form were ultimately, found in the laws of quantum mechanics, particularly in the principle that restricts the number of electrons that may occupy each orbit in an atom. Once these laws were understood, it was possible to predict the periodic table in a detailed way from fundamental physical principles. At present there is no scheme for generating the metabolic chart from such basics, but hope springs eternal. And maybe, just maybe, there is a missing law that will resolve the basis of biochemistry, just as the quantum mechanical principles predicted an explanation of the major features of chemistry.

In the absence of a basic principle, we search for unifying features, and they are not hard to find. Almost every sequence of metabolic processes involves one or more reactions with molecules of adenosine triphosphate (reporter from my youth, please note). This ubiquitous substance, best known by the abbreviation ATP, is central to energy processing in all cells. We find ATP printed along practically every line in the metabolic chart. If it were represented only once in our network of biochemical pathways, the drawing would be a giant rosette with all lines passing through the center. ATP is the final energy transfer molecule in almost all cellular processes. The reactions of this compound heat our bodies, power our muscles, charge our nerves, and otherwise drive the processes of life.

Accompanying ATP is a series of other substances that play major roles in energy transfer. Each contains the molecule adenine built into its structure. In the language of life this atomic configuration appears as the symbol for an energy storage molecule, yet the adenine portion itself plays no part in the energy process. The whole idea seems information-rich, somehow rather too linguistic or poetic for the grind-and-extract business of biochemistry; yet there it is. In addition to being a signal for energy transfer, adenine also constitutes a major symbolic component of the genetic code, being one of the four bases of DNA and RNA. Can there be some deep and fundamental, yet hidden, relationship between coding and energy transfer? It is a question worth addressing, for an understanding of adenine seems to lie close to the biochemical secrets of life.

Beyond the particular characteristics of the ubiquitous adenine, many less-sweeping propositions emerge from a study of the metabolic chart. These, too, stand as challenges to biophysicists, biochemists, and biophilosophers, urging them to penetrate deeper into the relations standing behind the experimental facts. Deriving the grand structure from a more primitive principle would give great insight into that perennial, yet ever significant, question: What is life? It would also spare another generation of students from having to memorize the entire chart.

Ice on the Rocks

For people who encounter ice most often in the form of cubes floating in a glass of diet soda or scotch and water, a large glacier is an awesome sight. The impact is doubled for those of us huddled on the cold, windy foredeck of the small cruise boat *Princess Patricia,* which at present is maneuvering carefully in the Tarr Inlet of Glacier Bay in Alaska. The Marjorie Glacier on the left, the Grand Pacific on the right, and a collection of assorted small icebergs floating by make it difficult to think of anything but water in its varied phases. It is, as it was for Coleridge's ancient mariner, a case of "water, water everywhere," but it is much more impressive when so much is in the solid form.

I imagine that humans living in today's interglacial age are not properly attuned to the presence of ice on the surface of the planet. We are intellectually aware of the vast stretches of frozen wastes in the Arctic and Antarctic and may even know that ice covers 7% of the planet's surface, but there is little direct feeling for a wall of compacted snow moving inexorably across the land. Eskimo cultures, with 12 or more words for snow, may have a deep appreciation for glaciers, but these objects are quite remote abstractions for residents of temperate climates. Only an occasional catastrophic event brings the great polar ice fields into public prominence. One such event occurred on the night of April 14, 1912, when the brand-new passenger liner *Titanic* tore into a large floating mass of ice that had calved from a North Atlantic glacier. The great liner sank rapidly, and the wandering mass of glacial debris continued its aimless journey until it finally melted away and also disappeared into the sea.

A taste of the next ice age is given us by Thornton Wilder in his enigmatic and highly successful play *Skin of Our Teeth*. Act I builds up to the lines, "Mr. Antrobus, what's that??—that big white thing? Mr. Antrobus, it's ICE. It's ICE!! . . . Will you please start handing up your chairs. We'll need everything for this fire. Save the human race. Ushers, will you push the chairs up here?" The rapid onset of the playwright's ice age may be an instance of dramatic license, but glaciers *can* move surprisingly fast. Many have been charted at one or two feet a day, with occasional bursts of higher velocity. The speed champ in my record book is the Black Rapids Glacier of Alaska, which reportedly raced up to 250 feet per day back in 1937. Glaciers recede as well as advance: The Grand Pacific displayed before our cruise boat has receded 65 miles up the bay since it was first discovered by the exploring Captain Vancouver in 1794.

Thornton Wilder is not the only literary figure to ponder civilization ending in an ice bath. The most imaginative modern treatment is Kurt Vonnegut Jr.'s *Cat's Cradle*, which is a combined science fiction story and morality play based on the imaginary "ice nine," a crystalline form of water that freezes at 114.4° Fahrenheit. A single seed crystal of this material is thus capable of solidifying all the oceans, bringing to an end life as we know it. Vonnegut's book is the most molecular of the ice age catastrophe epics, being based on the ways in which water molecules can aggregate into various states and thus introducing the fascinating domain of hydrogen bonds and crystal lattices. At the most fundamental level a glacier is, after all, a collection of very many H_2O molecules.

Water, composed of two hydrogen atoms and one oxygen atom, appears deceptively simple when studied in isolation. Yet when we read through scientific descriptions of its bulk properties, a single word keeps reappearing: *anomalous*. For example, when water freezes, it expands, in contrast to most compounds, which shrink to a greater density upon solidifying. This particular anomaly is what keeps icebergs afloat. Water has unexpectedly high freezing and boiling points. Water molecules also have an unusual capacity for sticking together, which produces not only the familiar ability of water to form beads on surfaces but a complex and variable structural arrangement among water molecules in the liquid state. Researchers have produced a long list showing that water has many other unusual properties when compared with other substances, especially those with similar chemical structures that led scientists to believe they ought to behave like water.

Thus water is both the commonest of substances and one of the most complex encountered by physical chemists. Many scientists have stressed that these curious features of water have all been used either in an ecological or molecular context to render the planet a suitable abode for life.

One of the more puzzling ways that living organisms capitalize on water's odd properties has to do with the control and use of energy, including electrical energy, inside cells. A few decades ago, when investigators began to study the electrical properties of pure water and ice, it was assumed that the electric current was carried by ions, electrically charged atoms or groups of atoms that move about the medium. Chief among these are positively charged hydronium ions of (H_3O^+) and negatively charged hydroxyl ions (OH^-). Since the movement of ions is much faster in the liquid state, it was further assumed that electrical conductance in the liquid would be much faster than in the solid. Once again water proved a surprise: ice was found to have an anomalously high conductance, comparable at 0° to that of liquid water.

If this is the stuff that science fiction stories are made of the actual explanation is equally novel. Although nearly all solids that conduct electricity employ small, fast-moving, negatively charged electrons, the charge carriers in ice are larger, slower, positively charged protons. Ice is, in fact, a proton conductor. So there are at least two ways in which electric current can flow through solids. A whole new world of protochemistry was opened, which is potentially as rich and interesting as the long-studied field of electrochemistry.

The loud bang of a small iceberg hitting the boat (the *Titanic* syndrome) triggers thoughts of the life history of glaciers. These huge structures have their beginnings high in the clouds, in microscopic centers of condensation that grow into snowflakes, beautiful crystalline objects that fall on the glacial surface, forming a light, fluffy aggregate. The fresh-fallen snow is immediately subject to compacting, crushing, local melting, and sublimation —the vaporization of substances directly from the solid state. The granules grow and pack, and the density rises. Heavy new snowfalls keep compressing the underlying material, squeezing out the air, and forming solid ice under substantial pressure.

At sufficiently high forces the bottom layer begins to flow because the bonds between molecules are deformed. Under the influence of gravity the entire glacier moves downhill across the rocky terrain toward the sea, where it eventually melts or breaks away at the extremities. A glacier grows, metamorphoses, and finally melts. The scale of events is hard for the chilled brain to comprehend: 10^{41} (one followed by 41 zeros) molecules of water sliding into the Tarr Inlet just off our bow.

Soon the *Princess Patricia* turns, and we begin to head down the bay. I cannot stop thinking about the relation of the macroscopic to the microscopic—hydrogen bonds and massive walls of blue ice. In the barroom we order a round of scotch and water to warm the bones. Observing a single floating ice cube reduces the magnitude of the problem to 10^{24} molecules of water. I slowly sip, and the miniature iceberg slowly melts. Just before it is too small to see, the cube is down to 10^{18} (one billion billion) molecules. There is great difficulty in bridging the gap between molecules and glaciers. Besides, my scotch is getting awfully watery.

Biphenyls and Bipeds

For the average bioscientist, visiting laboratories has a repetitious feel. After all, a centrifuge is just a centrifuge, the microscope hasn't changed all that much in the last 300 years, and the fundamental things apply as time goes by. But a day spent at the Oregon Regional Primate Center is, I must confess, different. The well-kept rhesus monkey enclosures set in the tranquil countryside give the institute visitor a feeling quite distinct from anything I had previously experienced.

Standing in the observation tower, as an object for the monkeys to stare at and chatter about, imparts the appropriate degree of humility. As one primate species to another, we both have reasons to be a little suspicious and very interested. Since we are in nominal control, the monkeys have the right to be more suspicious, and their actions reflect it. Another unique feature is seen by looking away from Mt. Hood in the background and gazing at a colony of ring-tailed Madagascar lemurs, one of the more exotic-appearing of the prosimians. It's good to know that this threatened species is alive and well in Oregon.

Primate centers have come to play an increasingly important role in modern biomedical research. Our views of the descent of man convince us that the animals housed in these centers are our near relatives and therefore very good models for the study of a number of normal phenomena and pathological states. In recent years the export of monkeys and apes from countries where most of the indigenous populations exist has been slowed down or stopped altogether by ecological or political concerns. The only way of ensuring a continuous supply of laboratory animals is to establish and maintain breeding colonies. This has been done in a number of centers around the United States. Most investigators strongly endorse the need for institutions to maintain stocks and carry out a variety of physiological and behavioral studies.

My host for today is Dr. Wilbur McNulty, an old friend from student days spent in the attic of Sloane Physics Laboratory at Yale, where we took part in constructing a do-it-yourself biophysics research facility and spent countless hours discussing the search for the secrets of life. The post–World War II period was filled with a warm, romantic spirit and deep faith in the powers of reason. Those of us who scientifically matured in that heady atmosphere are forever tinged with a glow of pride in having participated, however peripherally, in a great upheaval of human understanding. My friend and I have each moved on to different endeavors, yet we both fondly recall our early activities of focusing the cyclotron's deuteron beam on bacteria and viruses and assaying the results by whatever methods were available. Since a common intellectual heritage exists, resuming a discussion of scientific matters is quite easy even after a long hiatus.

The story that Bill is unfolding deals with the toxicity of 2,3,7,8 tetrachlorodibenzo-*p*-dioxin (TCDD) for rhesus monkeys. It all began in 1967, when "an epidemic disease with a high mortality struck rhesus monkeys kept for only a few weeks in newly constructed enclosures." The epidemic was finally traced to polychlorinated biphenyl (PCB) poisoning, and research on chlorinated hydrocarbons became a regular part of the center's program. *Macaca mulatta* proved to be a good experimental animal for this type of investigation, and careful experiments were carried out on the symptoms of PCB poisoning and the post-mortem findings on tissue changes at the microscopic level.

Subsequent work began on chlorinated dibenzo-*p*-dioxins, which must be among the most toxic materials ever created in the laboratory. Studies on the rhesus monkeys showed that the mean lethal dose was of the order of 2 μg/kg, or two parts per billion by weight. Reports from other workers indicated that guinea pigs were even more sensitive, with death of half the animals occurring at 0.6 parts per billion.

Thinking about toxicity data forces us back to the numerical question of how many molecules per cell would be required to kill organisms ingesting these very toxic compounds. Making some reasonable assumptions about the guinea pig results, we begin scratching our calculations on backs of envelopes (the romantic age predated computers) and come up with a value of about 30,000 molecules of TCDD per cell. Since a large fraction of material of this kind is absorbed in fats, the amount getting to cells is actually much lower than our average. And it seems reasonable to assume that toxic effects begin at appreciably smaller concentrations than the lethal dose. It is likely that a few hundred molecules of poison per cell cause substantial damage to sensitive organisms.

The most toxic material appearing in compendia is botulinus toxin, the lethal level of which is about one part per 10 billion by weight. Since the molecular weight of the protein that does the dirty work is very high, the effectiveness per molecule is even more extraordinary. Again back-of-the-envelope calculations reveal that only a few molecules of toxin per cell suffice to kill. The poison is so deadly because it targets functioning nerve cells and blocks the release of neurotransmitters. The invading molecules circulate in the bloodstream, locate the critical areas, and stick. Explaining the efficacy of the botulinus agent may provide a line of thought for trying to understand why industrial organics, such as the PCBs and more especially the TCDDs, are so frighteningly toxic.

One possibility is that the highly dangerous compounds resemble some normally occurring substance and bind to critical targets on or within cells. Sitting on these sites, the poisons might inhibit or block some vital biological function. But what normally occurring molecules could PCBs and TCDDs resemble sufficiently closely to confuse the cellular recognition mechanisms? This question sends us scurrying for biochemistry books. Sure enough, we find that the hormone thyroxine is a tetrahalogenated biphenyl. Rather than containing four chlorines, like TCDD, thyroxine contains four iodines, but there is a lot of similarity in their atomic structures. We do, of course, find the intriguing sentence, "The precise method by which thyroxine stimulates oxygen consumption of the tissue is not known" (A. Lehninger, *Biochemistry*, 1976). It is therefore possible that the reagents are inhibiting some ill-understood reactions controlled by thyroxine. Or do there exist other as yet undiscovered hormones that involve a similar molecular structure of two benzene rings and a number of iodines or chlorines? We are off into the speculative realms, and the activity recalls the exciting discussions of graduate school days.

The problems are not solved by brainstorming alone, and we realize that they will yield only to experiment and more experiment. I leave Oregon with a fistful of reprints and the opportunity to further learn of the results in rhesus monkeys. I leave also with a deep appreciation of the work being carried out in these laboratories. The experiments on PCBs and a wide variety of other compounds provide some of the best information we have with which to approach establishing standards for humans. "Reasonable standard" is, however, a difficult concept. If one part in a billion is toxic and a reagent accumulates in the body, then what is the safety level of food and water? Is it one part in 10 billion or one part in 1,000 billion? We are beginning to get to homeopathic amounts, wherein lies much of the monkey business taking place in establishing environmental standards. But that's another story, and for today I'm satisfied to enjoy the euphoria of playing with novel ideas and recapturing the fun of those days when everything was new. Come to think of it, there's a lot to learn, and the intellectual excitement of youth never loses its thrill.

Living Lodestones

Just when it seems we have mastered almost everything there is to know about bacteria, some new and totally unexpected phenomenon comes along to renew our respect for the subtlety of living processes even at the level of these tiny prokaryotes. Richard Blakemore's discovery (*Science* 190:377, 1975) of microorganisms that preferentially swim north along the lines of the earth's magnetic field is just such a chastening happening. I doubt that any of us would have predicted the existence of these creatures or the manner in which they operate. The latter has now been made clear in some fruitful collaboration between Blakemore and physicist Richard Frankel of the National Magnet Laboratory.

To place matters in appropriate historical context, recall that magnetism was one of the earliest subjects of scientific study in both the Orient and the West. The oldest recorded writings on the subject come from ninth-century-B.C. Greece and deal with magnetite, a mineral that in its normal state is strongly attracted to iron. This naturally magnetic material was mined in the province of Magnesia in Thessaly and was described successively by Thales, Plato, and Lucretius. The navigational compass of the Chinese may have predated the Greek discoveries, but records are uncertain. Writings from A.D. 120 mention the directive power of magnets and suggest that the compass had long been in use by travelers across the plains of Asia.

A scientific investigation of natural lodestones was reported in a thirteenth-century manuscript by Petrus Peregrinus de Maricourt. These early studies reached a high point in the epic work *Epistola de Magnete*, published by William Gilbert, personal physician to Queen Elizabeth I. Gilbert established that the earth is a gigantic magnet whose attractive poles closely correspond to the geographical ones. He understood the phenomenon of dip (the downward pointing of the north-seeking pole in the Northern Hemisphere and conversely in the Southern) and devised an instrument to measure it numerically.

Since Gilbert's classical work there have been numerous reports of organisms responding to the earth's magnetic field, but some ambiguity has surrounded many of these studies, and no mechanism of interaction of organism and field has been established. The bacterial studies have now provided definite evidence of that interaction.

Magnetotactic microorganisms have been collected from marine and freshwater sediments and normally live in low oxygen concentrations. Their swimming motion is linear, compared with most bacteria, which periodically tumble and change direction. Each bacterium is propelled by flagella. With respect to orientation, cells act as single magnetic dipoles; that is, their behavior is the same as compass needles or bar magnets. If an external magnetic field is applied, the long axis of the cell parallels the lines of force. If the field is rotated, the axis follows. Bacteria clearly discovered the principle of the compass eons before their distant human relatives managed to accomplish that feat.

When examined by electron microscopy, magnetic bacteria show one or two rows of small, dense particles running along the cell axis. The size of these granules is characteristic of the species, ranging about a tenth of a micron across. They contain most of the cell's iron. In order to determine the chemical form of the iron atoms, studies were undertaken with Mössbauer spectroscopy, a technique especially sensitive to the chemical state of certain metals in crystals. The spectra showed that the particles are single-domain magnetic crystals, the same kind of material that the peasants of Thessaly had found as attractors of iron. From a knowledge of the material and the size of the chains, it is possible to calculate the dipole strength and confirm that the earth's field is strong enough to orient these bacteria. The magnetite in these cells constitutes a biomagnetic compass.

The first magnetic bacteria to be studied came from the northern half of the globe and uniformly swam to the north magnetic pole. Since they are propelled by asymmetrically located flagella, there is a required relation between dipole and flagella orientations that must be maintained on cell division. In addition, these organisms must be able to synthesize single-domain magnetite crystals of an appropriate size—a bit of solid-state technology that seems particularly impressive and might have some commercial application.

Having discovered the remarkable property of magnetic bacteria in Massachusetts and other parts of the United States, Blakemore and Frankel were directed by the geomagnetic logic of our planet to Australia and New Zealand, where they made the symmetry-satisfying observation of similar bacteria swimming toward the South Pole. This provided a clue to the biological raison d'être of magnetotaxis, which takes us back to Dr. Gilbert and his dip meter. In the Northern Hemisphere the field is directed downward, for a north magnetic pole, so that a north-swimming bacterium has a component of motion directed down into the sediment. If such a bacterium were taken to the Southern Hemisphere, it would swim upward, away from its food and anaerobic habitat. In the Southern Hemisphere the polarity is reversed, and indigenous south-seeking bacteria are also directed to their optimal habitat. The major biological role of magnetotaxis, therefore, appears to be keeping these cells in or near the bottom sediment.

There seems to be a generalization emerging in science over the last 100 years or so that whenever a principle of physics exists that is applicable within the constraints of organic life, then some species or higher taxon has developed a way to utilize that law of nature in the competition for survival. This excludes phenomena that might require very low or very high temperatures or high vacuum but includes almost everything else. An instructive example is the way a number of very subtle principles of aerodynamics have been employed in the flight of birds. Another case in point is the application of surface tension by insects that live their entire lives at the air-water interface. The same type of surface tension coupled to the tensile strength of water is used for transporting fluid to the tops of giant sequoia trees. The interaction of chemical and osmotic transport in kidneys provides further appreciation of our generalization. The extremely sensitive mammalian transducers of sound, light, and chemical concentration (taste and smell) demonstrate applications of physics that we still don't completely understand. We are now beginning to learn of a remarkable array of essentially solid-state devices, such as membrane-bound enzymes, transmembrane ion transporters, flagella motors, and single-domain magnetic crystals of rather uniform size. In three billion years of evolution there seems to have been a thorough sampling of all mechanisms that can be built out of proteins, lipids, carbohydrates, and nucleic acids.

Returning to the subject of magnetite crystals, we know that they are not restricted to bacteria but also have been reported in the tongue structure of certain mollusks, the abdomens of bees, and the skulls of pigeons. In bees and pigeons it has been postulated that the magnets may be part of navigational devices. Although the role of magnetite in these higher organisms is not certain, it's safe to assume that a function exists and we are due for even more surprises in the field of animal magnetism.

Perhaps the most novel aspect of the bacterial story relates to the fact that we have always considered the microbial habitat in terms of local variables in the immediate neighborhood of these tiny cells. Now we find ecologically significant interactions between bacteria and the earth's magnetic field, the most global of all parameters. This suggests that we are again being treated to an example of the relation between the microscopic and macroscopic domains.

From Soup to Solid-State

The concept of energy, the very core of our understanding in physics, is sufficiently elusive in biology that we continually struggle to extend its applications. It requires both discipline and effort to deal with the fact that every breath we take is part of a cosmic transformation that began with the "big bang" and will end either in the heat death of the universe or in some regenerating catastrophe. Yet there is no doubt that our muscles are powered by chemical potential derived from sunlight, which itself results from thermonuclear reactions in the interior of our star. On a more earthly plane, the last 20 years have witnessed a radical and exciting reassessment in our view of cellular processing and storage of that most precious of all commodities: energy.

Our background story begins in 1941 with Fritz Lipmann announcing the concept of high-energy phosphate bonds as the ubiquitous energy-intermediate in biology. Muscles, fireflies, electric eels, and biosynthesis are all powered by energy temporarily stored in molecules of adenosine triphosphate (ATP), the universal medium of power transfer. ATP constitutes a biological storage unit, which can exist in an "energy-rich" state with the intact phosphate bond or in the discharged state, where it is broken down to adenosine diphosphate and free phosphate. In the discharge process, energy is transferred to cellular hardware to do the work of life. In animal cells almost all of the energy used for ATP synthesis comes from the oxidation of sugars or other storage molecules by atmospheric oxygen. The ATP theory has provided a firm basis for biochemistry and has been reinforced, elaborated, and experimentally confirmed since its original statement. The Lipmann formulation left open the question of how the high-energy compound was made from its component parts.

During the 1930s, when Lipmann's work was being carried out, a dichotomy of method existed between biochemistry and cell biology. Biochemists ground up tissues to extract enzymes and substrates, purify them, and study them in the laboratory. Cell biologists were interested in the basic unit of life as a collection of interrelated organelles best studied in intact structures. The latter accused the former of regarding the cell merely as a bag full of molecules operating as if they were in a test tube, and there was some justification to their criticism. Therefore it was not strange that those with a chemical orientation assumed that ATP was synthesized by normal solution chemistry—in the soup, so to speak.

Over the next 20 years it became apparent that the mitochondria, tiny intracellular organelles, were the sites of most ATP production in all higher organisms. It was found that intact mitochondria with unbroken membranes were required, an unexpected result from the point of view of ordinary chemistry. This result drew cell biology and biochemistry closer together, because it was conceptually not possible to separate structure from function in this problem. The two disciplines were also associated by developments in electron microscopy, which allowed biologists to look at entities much nearer in size to the molecules of the chemists than had previously been possible, further blurring the structure-function dualism.

As to the chemical processes, there was a strong consensus that the oxidation produced a high-energy chemical intermediate that was then linked to the synthesis of ATP. Two difficulties made this consensus shaky: No one could find the intermediate, and there was no reason why such a chemical substance should require an intact organelle in order to react.

At this point the Mitchell chemiosmotic hypothesis appeared on the scene It postulated that oxidation-reduction reactions moved protons from one side of the functional mitochondrial membrane to the other; the energy was stored in this form of charge separation and then used to synthesize ATP. It is hard to look back over even these few years and imagine how radical the hypothesis seemed to many scientists at that time. I have been told that when Dr. Mitchell presented a talk on his concepts in Amsterdam in the early 1960s, there were those in the audience who found his views so hard to comprehend that they questioned his mental stability.

The theory was strange in a number of ways. It associated a chemical reaction (oxidation-reduction) with the directional movement of ions. To the old "grind and extract" school there was no spatial direction to chemical processes; they took place in solution. The fact that membranes could organize chemical reactions in an asymmetric way had just not been given very much attention. The mode of storage also was unusual: Separating electrical charges across a membrane was similar to charging up a capacitor in electricity; what was so different was that all previous capacitors used electrons, while this one used protons. Protochemistry, as distinguished from electrochemistry, was virtually unknown in 1960, although it was theoretically well within the realm of possibility. Third, it was not known how the separation of protons across a membrane could be coupled to ATP production. This last feature has only recently been shown, although the thermodynamic possibility was long recognized.

The years following the introduction of Mitchell's views were characterized by vigorous controversy, as a number of more or less detailed hypotheses were put forth to explain how the energy of oxidation ended up in ATP. An enormous amount of experimental work was carried out by groups of researchers all around the world, and each new piece of data seemed to be explicable by all of the models. Results gradually came in from related fields, such as chloroplast photosynthesis, ion transport in bacteria, photoenergy conversion in the salt bacteria, and bacterial motility. As the aggregate of these experiments and the connections between them became understood, more and more scientists began to see the energy of charge separation using protons as a unifying theory for much of bioenergetics. In 1978 the Nobel Prize was awarded to Peter Mitchell. From amused amazement to the accepted establishment view took less than 20 years, often difficult years for the protagonists.

The current view of bioenergetics has fully embodied the concept that the separation of hydrogen ions is the intermediate energy storage between oxidation-reduction steps and the making of ATP. When it comes to exploring the detailed molecular mechanisms by which these remarkable reactions occur within a membrane, we are once again led into new devices that seem to come under the heading of solid-state physics and chemistry. The protein molecules embedded in the membrane appear to function as tiny transducers that can transport and convert the potential of protons into the chemical energy of ATP. The task before us is to learn how these microelements work. This will lead to a better understanding of energy functions in living cells, and tricks learned from nature undoubtedly will find industrial applications. Chemists have basically mastered the coupling of electricity and chemical reactions through electrons; the mitochondrial method of doing it is through protons. The results of this discovery have not been fully explored, and when they are worked out, we anticipate the emergence of some exciting novel technology and new questions to study. It may even teach us something about physics.

Entropy Anyone?

To judge from the writings of C. P. Snow, entropy and the second law of thermodynamics were once indelicate subjects. Now things have changed, and on the cocktail party circuit we hear of entropy in art, entropy in economics, entropy in urban decay, and other erudite-sounding applications. A most difficult concept in physics is being applied to confused areas in the social sciences, with the impression being conveyed that this increases our comprehension. Clearly, such a situation should send us scurrying back to the first principles of physics to clarify the issues.

Thermodynamics, the science of heat and work, was firmly established in the mid-1800s. Its development paralleled the Industrial Revolution, an era characterized by the application of heat engines in manufacture and transport. Studying the efficiency of these machines was the task of science, although generations of aspiring engineers have repeated the adage that the "steam engine has done more for thermodynamics than thermodynamics ever did for the steam engine." During the decade from 1840 to 1850, two physicians, Mayer and Helmholtz, developed a full understanding of the concept of energy and derived the conservation law. The principle that energy can be changed from one form to another but can never be created or destroyed has been an enduring cornerstone of physics.

Using this new knowledge, Rudolf Clausius began his studies of the efficiency with which heat could be transformed into work. He set down the second law of thermodynamics in the form "heat cannot of itself pass from a colder to a hotter body." From this experimental generalization and the work of Lord Kelvin he was able to deduce that engines could not be 100% efficient; it was impossible to transform all the heat into work. In seeking a formal measure of the energy unavailable as work, Clausius, in 1865, introduced the word *entropie*, derived from the Greek root *tropie*, or "transformation." This parameter, in classical theory, permits a determination of the amount of energy that cannot be recovered as mechanical work.

The precise Clausius definition of entropy was abstract, mathematical, and formulated in terms of differential calculus. It provides an excellent way of carrying out calculations but yields very little in terms of intuitive notions. In this branch of physics we use such quantities as pressure, volume, and temperature, which can easily be related to everyday experience. Other parameters, such as work, heat, and energy, are harder to conceptualize but make contact with familiar ideas. Finally, we have quantities, such as entropy, that are almost entirely dependent on abstractions and therefore are the most difficult to understand.

In developing their science, the founders of thermodynamics also introduced and stressed the concept of equilibrium to provide reference states upon which a comprehensible theory could be based. A system at equilibrium must have aged for a long time in isolation from active processes. This was done by immersing the system of interest in a heat bath or placing it in a rigid insulated box; under these conditions, it exchanges neither materials nor energy with the rest of the universe. Such collections of matter are basically dead and uninteresting. They can be investigated in the laboratory but rarely exist in the real world. Equilibrium is an idealization that has enabled scientists to gain a deep and perceptive insight into many natural processes; however, such measures as entropy, which have a precise meaning only at equilibrium, must be used very cautiously in the far-from-equilibrium everyday world. Using these ideas, Clausius and Lord Kelvin recast the second law of thermodynamics in the statement that for spontaneous processes entropy always increases. This was widely interpreted as meaning that the universe was running downhill toward a "heat death." It was assumed that physical processes always led to the most disordered states.

Nineteenth-century biologists, fresh from the triumphs of the theory of evolution, were baffled by the seeming contradiction between the degradative thrust of the second law and the progressive direction of evolution toward higher organization. They wondered if life, with all its anabolic processes, was a violation of the known principles of physics. The seeming paradox was first resolved by the perceptive Ludwig Boltzmann, who pointed out that the surface of the earth and the biosphere were kept far from equilibrium respectively by the constant flow of sunlight and by continuing photosynthesis. Such systems under an energy flow were not constrained to undergo an increase in entropy but could indeed experience a decrease to more ordered states. Life was fully consistent with the laws of thermodynamics.

During the following period an elaborate structure of thermal physics and chemistry emerged and helped to provide an understanding of our world. Boltzmann, Gibbs, and Einstein demonstrated that laws of thermodynamics are exact because ordinary measurements are averages over huge numbers of atomic and molecular events. Although the individual behavior of these fundamental entities is very hard to predict, their statistical properties can be precisely accounted for.

In the years from 1880 to 1915 the atomic theory became universally accepted by the scientific community. One necessary characteristic of all matter is the constant random motion of atoms and molecules when viewed on their own size scale. The glass of water serenely sitting on my desk consists of 10^{24} molecules, which are constantly moving about, dissociating into hydrogen and hydroxyl ions and then reassociating, forming and unforming microcrystals, and colliding with the walls of the container. Because the individual molecules are so small, we see only averages over vast numbers of particles and hence the illusion of great regularity.

Underlying this external impression of order is a real molecular chaos that has been studied by statistical mechanics and kinetic theory. One of the major triumphs of those disciplines has been a development of the entropy function as a quantitative measure of that submicroscopic disorder. The reason why some of the heat energy is unavailable for work is that it is randomly distributed among molecular modes and cannot be extracted in any coherent way by large-scale devices. This gives added insight to Clausius's finding. With the advent of information theory came the realization that entropy relates to disorder by providing a measure of our residual ignorance about the molecular details of a system after we have acquired all the necessary quantitative data at the size range of ordinary instruments.

Returning to the modern popularizers of entropy as a universal explanatory, I have before me a letter from Jeremy Rifkin, author of *Entropy: A New World View*. His communication states that "entropy helps explain why we have runaway inflation, soaring unemployment, bloated bureaucracies, a widely escalating energy crisis, and worsening pollution." That sentence is an archetypical example of the patent intellectual nonsense being offered in applying physics to the social sciences. Entropy, a deep and hard-to-penetrate physical construct, is being indiscriminately applied to situations in which it is devoid of physical meaning. No hint is given that the far-from-equilibrium world in which we live is governed by different laws than the thermodynamics learned in an introductory physics course. Boltzmann's work of a hundred years ago is ignored, as are numerous works of modern science showing the antientropic character of energy flow processes. In short, entropy is being made a code word to explain every political idea Rifkin wishes to propound. This is a dangerous misuse of science, because it may fool the scientifically untrained into believing that certain conclusions come from natural laws, when in fact they do not. Fooling the public with "sciencese" decreases our society's ability to make rational judgments.

The popular purveyors of entropy pap should be reminded of the words of Pope:

> *A little learning is a dangerous thing;*
> *Drink deep, or taste not the Pierian spring;*
> *There shallow draughts intoxicate the brain.*
> *And drinking largely sobers us again.*

In the meantime, what of the notion of entropy? This elegant quantity is a precisely defined construct of physics; it can be rigorously measured for equilibrium systems and can be given meaning for near-to-equilibrium considerations. The second law, when formulated in terms of entropy, gives powerful insights into a wide variety of problems. It seems a shame to take such a beautiful and exact idea and blur its meaning by indiscriminately applying it to all sorts of areas that have nothing to do with equilibrium thermodynamics. Such a procedure might disorder our thoughts about other disciplines that are already difficult enough.

The Odd Couple

Molybdenum, at first glance, seems like an unlikely candidate to play a vital role in the living world. It occurs naturally mostly in sulfur-containing minerals or in combination with lead and oxygen in the ore wulfenite. The pure substance is a silvery-white, very hard metal that appears far removed from organic forms. However, in spite of these impressions, biochemical research has now proved beyond any reasonable doubt that this heavy element is an absolute requirement for the functioning of large-scale ecological systems. In addition, several molybdenum-containing enzymes have been identified and characterized.

Far removed from molybdenum in the periodic table and different from it in every conceivable property is the element nitrogen, which occurs most abundantly as a rather unreactive fraction of the earth's atmosphere and is a vital atomic component of living systems. It is one of the big six (carbon, hydrogen, nitrogen, oxygen, phosphorus, and sulfur) that make up well over 95% of all living material. This low-atomic-weight element is necessary for the structure of proteins and nucleic acids, the major functional and information-carrying molecules of the living world. Several ecologically essential chemical reactions of nitrogen compounds are catalyzed exclusively by molybdenum in biological molecules. Thus, this unlikely pair of elements, nitrogen and molybdenum, are forever bound together in their biological roles.

To understand this odd relationship between unlike elements in more detail, we need to examine the nitrogen cycle, that very large flow of matter from atmospheric molecular nitrogen, N_2, to amino and nucleic acids and once again back to the atmosphere. Nitrogen atoms can occur in a wide variety of oxidation states, from the least oxidized form of ammonia to the most oxidized form of nitrate. Nitrogen gas lies halfway between. The oxidation state is determined by the number of electrons associated with an atom. The transfer between these more or less oxidized forms constitute the chemical steps that make up the nitrogen cycle. Certain crucial stages in these reactions require active sites containing molybdenum atoms.

Among the biochemical reactions involving nitrogen, fixation of that element is best known because of its vital role in agriculture and the global ecological economy. In this process gaseous nitrogen molecules are taken up by living cells and converted into the biologically useful form ammonia. Since subsequent stages of the cycle result in N_2 being given off by living material and returned to the air, fixation is necessary lest all of the nitrogen be lost from the living world and end up in the atmosphere. In the balance of nature, all biological nitrogen credits come from fixation. Were it not for these deposits, the withdrawals due to denitrification would deplete the reservoir, and life would slowly grind to a halt. Thus, agronomists are constantly adding fixed nitrogen to soil. This material comes from either naturally occurring guano reserves or the industrial synthesis of ammonia and nitrate.

In the world of nature, capturing gaseous nitrogen is the province of a few types of simple microbes. The best known of these are bacteria of the genus Rhizobium, which live in symbiotic association with the roots of leguminous plants, such as soybeans, peas, and alfalfa. The nitrogen-fixing organisms are found in nodules, where they extract energy from the plants' sugars and pay their way by providing their hosts with available nitrogen. Another group of free-living soil organisms also can perform the fixation trick, as can several blue-green algae that inhabit lakes, rivers, oceans, and damp, sunny places. There are fewer than 100 known species that can accomplish the chemical process of fixing nitrogen. The entire biotic world is in their debt for its very existence. I can think of no small group of species whose extinction would be more devastating.

Every microorganism capable of fixing nitrogen produces the enzyme nitrogenase, a high-molecular-weight molecule containing both iron and molybdenum. The metalloprotein has been isolated from a number of species and appears to be virtually the same molecule regardless of its source. The discovery of a cellular mechanism for converting nitrogen to ammonia must have occurred only once in the history of the planet, and all of the subsequent availability of biologically usable nitrogen follows from that event. In all known cases molybdenum is absolutely necessary for the biological fixation of nitrogen.

Having listed the experimental facts, we are led to the question: Why molybdenum? It is not very abundant on the surface of the earth. Crick and Orgel have suggested that the obligatory requirement for such a rare material might argue for life's having migrated to earth from some place in the universe where it was more abundant. The alternative argument is that the unique properties of this element are of sufficient value that its extraction, even from low-concentration environments, is highly advantageous. These evolutionary arguments would explain the survival and spreading of genes for molybdoenzymes.

Molybdenum is one of the transition elements, a large group of metals made up of atoms that are able to exist in many oxidation states. When such an atom loses an electron, it becomes more oxidized; when it gains an electron, it becomes more reduced. Such metals are generally used by biological systems in enzymes that carry out oxidation-reduction reactions. Iron, copper, and molybdenum are the main atoms employed in electron-transfer reactions of biochemistry.

Since nitrogen (N_2) is so unreactive, it requires some very special conditions in order to be reduced to ammonia (NH_3). Industrially this is done by mixing nitrogen and hydrogen gases at 500° C and 1,000 atmospheres of pressure in the presence of an iron catalyst. Realizing the extreme conditions that must be used in the commercial synthesis leads to an appreciation of the subtleties that are employed by the molybdenum-containing enzyme molecules in carrying out the reaction under the mildest of conditions of temperature and pressure. E. I. Stiefel and his co-workers have suggested that molybdenum among all the transition elements may be specially suited to the role of bringing electrons and protons to and from nitrogen atoms. If this turns out to be the case, then the odd couple will turn out to be associated by very basic thermodynamic and quantum-mechanical properties inherent in their structures and positions in the periodic table.

There are still a number of mysteries to be cleared up before we fully understand the nitrogen-molybdenum coupling. The study of molybdenum has some special difficulties, among which are learning to spell the word once you pronounce it or to pronounce it once you see the spelling. However, scientists are persistent types and doubtless will overcome this difficulty as well as others in comprehending the full and seemingly unique biological role of molybdenum atoms.

Energy Flow in Ulysses

While James Joyce is universally regarded as one of the literary lights of the twentieth century, his relation to the sciences has not been stressed. He did try attending medical school in Dublin with the thought of making a living as a doctor while following his calling as a writer, but after a few lectures he abandoned the enterprise and left for Paris. Nevertheless, for an intellectual giant like Joyce, all human knowledge is appropriate subject material, and strange bits of natural philosophy keep appearing throughout the pages of *Ulysses*.

I must at the outset confess that, mea culpa, I have never read all the way through *Ulysses*, so the final few hundred pages are for me a mystery of considerable depth. Thus, it happened the other day that while trying to track down a quotation on water appearing near the end of the book, I stumbled upon a fascinating paragraph on my favorite subject, thermodynamics. This relation between water and fire was not accidental, occurring in the story at 2:00 A.M., when Leopold Bloom was setting a kettle of water to boil. The section on heat begins,

> What concomitant phenomenon took place in the vessel of liquid by the agency of fire?

That question is followed by quite detailed scientific description embedded in the richness of Joycean prose. In the spirit of increasing the communication between the sciences and the humanities, I present a scientist's annotated version of an artist's view of science.

The answer to what concomitant phenomenon took place starts with

> The phenomenon of ebullition.

This, of course, translates as "boiling," but the formal style announces the erudition to follow:

> Fanned by a constant updraught of ventilation between the kitchen and the chimneyflue, ignition was communicated from the faggots of precombustible fuel to polyhedral masses of bituminous coal.

Faggots refers to bundles or pieces of kindling wood. *Precombustible* is a strange usage, since all fuel is combustible. The implication is that the faggots must first combust to initiate the main process or burning of the coal. The paragraph continues,

> containing in compressed mineral form the foliated fossilised decidua of primeval forests which had in turn derived their vegetative existence from the sun, primal source of heat (radiant).

In this part of the sentence Joyce both reviews the process of coal formation and acknowledges the sun as the source of biological power. Here and subsequently he makes full contact with the energy flow view of biology, which is so prominent in contemporary ecology. His use of *decidua* to describe the annual leaf crop and remains of deciduous plants is unrecognized by the *Oxford English Dictionary*, which defines a decidua as a uterine membrane.

The author does seem to have firm knowledge that the energy we receive from the sun is in the form of radiation that is

> transmitted through omnipresent luminiferous diathermanous ether.

For 19th-century physicists the ether was a weightless substance that permeated all matter and space and was the medium for the transmission of electromagnetic energy. Its complete transparency is described by the words *luminiferous* and *diathermanous*. Also, by the time *Ulysses* was published in 1922, the concept of ether had long been abandoned as a result of Einstein's special theory of relativity, published in 1905. Artists do face problems with the constantly changing nature of the scientific description of the world. Nevertheless the language here is highly sophisticated and highly specialized scientific jargon, and one does wonder where Joyce would have encountered it.

> Heat (convected), a mode of motion developed by such combustion, was constantly and increasingly conveyed from the source of calorification to the liquid contained in the vessel.

In this sentence the author expresses the second law of thermodynamics in terms of the flow of heat from a hotter body to a colder body. There also seems to be a recognition that heat is a form of molecular motion. *Calorification* is a rather ornate word for heat production, but this paragraph seems to abound in ornate words. The sentence goes on,

> being radiated through the uneven unpolished dark surface of the metal iron, in part reflected, in part absorbed, in part transmitted, gradually raising the temperature of the water from normal to boiling point.

This is a simplification of the complicated processes by which energy is transferred from the glowing coals to the pot. It does recognize that the processes are complex but seems to ignore conduction as a method of heat transfer. The description does imply knowledge that dark objects are good absorbers of heat. Joyce's knowledge seems extensive, but not complete, as if he is reaching slightly beyond his grasp in including such detailed science.

The next clause is,

> a rise in temperature expressible as the result of an expenditure of 72 thermal units needed to raise 1 pound of water from 50° to 212° Fahrenheit.

This part of Joyce's thermal physics turns out to be the most troublesome of all because it involves numerical values that appear to be incorrect. To see the problem, we first must inquire as to the meaning of *thermal units.* One would assume that the reference is to what is conventionally designated a British thermal unit, but having been educated in Dublin, Joyce learned the term without the prefix *British*, which would have been an unnecessary insult to an Irish physics class. Such a unit, the BTU, is the amount of energy required to raise the temperature of one pound of water by one degree Fahrenheit. Heaters and air conditioners in the United States are still rated in BTUs.

If Joyce was referring to this kind of thermal unit, then 162 thermal units would have been required to raise a pound of water to the boiling point, not the 72 units stated. Either the author has made a mistake, or he is using a thermal unit unknown to me, or he has invented a new unit, or he has a hidden message. For example, 72 can be derived from 162 by adding the one and the six to get seven. Were the example from the cryptic *Finnegan's Wake*, I would have opted for the esoteric explanation. As it is, I'm inclined to think that the great literary genius, who sometimes revised his manuscripts 14 times, was just careless with his arithmetic or a typographic error has crept in. The "50°" has previously been referred to as normal, or what I would assume was room temperature. The hearty Dubliners were not given to overheating their houses at night.

> What announced the accomplishment of this rise in temperature?
>
> A double falciform ejection of water vapour from under the kettlelid at both sides simultaneously.

Falciform means "sickle-shaped, curved, hooked." Joyce was clearly a man who loved words and left no vocabulary unexpressed as he labored over his manuscripts. He goes on:

> For what personal purpose could Bloom have applied the water so boiled?
>
> To shave himself.

We have come a full course from the simple act of setting a kettle on the fire to the simple act of shaving. In between is found a complex artistic description of thermal physics along with Joyce's insights and mistakes. Literary greatness takes strange forms.

Having annotated the section on fire, I am tempted to go back to analyze the much longer section on water that I was originally seeking. But the task seems too overwhelming. The long discussion begins with Roundwood reservoir in county Wicklow and ends with stagnant pools in the waning moon. In between is a Niagara of choice Joycean verbiage. In any case, the message is clear: I must return to the reading of *Ulysses*. Oh, when will I ever get time for backgammon?

The Beauty of Mathematics

In an age dominated by scientific technology there is a continuing problem of communicating the content and meaning of science to a broader public. Researchers still find it difficult to translate the latest technical perspective into language that is understandable to nonspecialists. It often takes 50 years or more for novel theoretical ideas to become incorporated into the thinking of the greater intellectual community. The viewpoint of quantum mechanics, for example, put forward in the late 1920s, has still not had its full impact on humanistic thought. The material is that hard to digest. Yet there are few philosophers of science who doubt that this theory has and will continue to exert a deep effect on our views of man and nature.

The one aspect of science that is most enigmatic and difficult for outsiders to comprehend is the use of mathematical abstractions to describe events that are very concrete. A physicist sits at his desk writing a series of symbols and numbers, while out in space a planet orbits the sun. Exact instruments are used to take very careful measurements of the planetary positions, and *voilà*, the numbers obtained agree with the pencil-and-paper calculations. The whole space program, for example, depends on our ability to calculate exact orbits. Scientists' familiarity with this correlation between theory and experiment dulls us to the strangeness of the relation between the observed world and the little marks we put on paper.

There is a beautiful experience, available to physics students in their sophomore year of college, that illustrates the emotional impact of being able to deal with predictive mathematical theory. The first step in this exercise is to set down on a piece of paper in mathematical form Newton's second law of motion, which states that force equals mass times acceleration. Next, one writes the equation that means there is a universal force of gravity, an attraction between all bodies that is proportional to their masses and inversely proportional to the square of the distance between them. The final step is to carry out a few lines of mathematical manipulations that quickly yield equations for Johannes Kepler's laws of planetary motion.

The act of personally deriving the motion of the bodies of the solar system from a few simple assumptions is a profound experience. When the result emerges, there is in the minds of many students a sense of *déjà vu*, the feeling of a return to some primordial knowledge.

Some scientists greeted with awe the act of predicting the future of celestial systems. Others responded with arrogance. After Pierre-Simon Laplace's *Celestial Mechanics* was published, for instance, the scientist met with Napoleon, who noted that he saw no mention of God in the book. Laplace is reported to have replied, "I have no need of that hypothesis."

The most noteworthy achievement of mathematics in celestial mechanics did not occur until more than 150 years after Newton's laws of motion and universal gravitation were published. Astronomers studying the planet Uranus were unable to explain its motion by the laws of mechanics. They postulated that the orbit was being influenced by an unknown planet farther out from the sun. Astrophysicists used the mathematical methods of Laplace to predict the path of the unknown object. This information was communicated to observers who in 1846 aimed their telescopes in the predicted direction and discovered the planet

Neptune. This discovery, one of the great triumphs of the human intellect, was totally dependent on the curious relation of the abstractions of mathematics to the concrete world of physics.

Once one understands and feels the power of the derivation of Kepler's laws and subsequent developments, he may become more mystical about mathematics or more committed to hard-nosed materialism. But one way or another the mathematical analysis is a moving experience.

Throughout physics, chemistry, and parts of biology, this strange relation between equations and real-world events keeps appearing. The scientist uses mathematics because it gives him both an abbreviated way of representing experience and a feeling of deep understanding. Nevertheless there remains a whimsical wonder about why mathematics works. Theoretical physicist and Nobel laureate Eugene Wigner once captured this feeling in an essay entitled, "The Unreasonable Effectiveness of Mathematics in the Natural Sciences."

Wigner's first argument begins with a statement of Einstein's that "the only physical theories which we are willing to accept are the beautiful ones." Mathematics seems to be the only study that generates sufficient beauty for the physicist. This tells us more about physicists than about physics, but in Eugene Wigner's view of science the subject material is, in any case, inseparable from the mind of the scientist.

After giving several examples of the applications of very abstract mathematics to specific problems of atomic structure and spectra, Wigner comes to the conclusion that "the unreasonable effectiveness of mathematics" in physics is an empirical law of epistemology. Somehow the structure of mathematics and the structure of the universe are close enough that one may represent aspects of the other with great precision.

Wigner's essay concludes with the following:

> Let me end on a more cheerful note. The miracle of the appropriateness of the language of mathematics for the formulation of the laws of physics is a wonderful gift which we

> neither understand nor deserve. We should be grateful for it and hope that it will remain valid in future research and that it will extend, for better or for worse, to our pleasure, even though perhaps also to our bafflement, to wide branches of learning.

Thus the scientist uses this wonderful gift of mathematics but, like the layman, remains puzzled as to why it works. Knowing that even the experts wonder will perhaps lessen the fear and loathing of mathematics on the part of so many people outside science. To some of us, equations are objects of beauty, but to all of us they are tools of the trade.

A difficulty in coping with the mathematical side of science is worth overcoming if one wishes to be fully sympathetic with contemporary scientific thought. Even a rudimentary knowledge of mathematics will go a long way toward giving one a feel for the language of natural philosophy.

INDEX

Aboriginal beings on Galápagos, 4
Acropolis, 133–34
Adenosine triphosphate (ATP), 197–200, 214–17
Afterlife
survival of the soul, 101
replaced by longevity, 102
Against Method by Paul Feyerabend, 111
Age restrictions, 68–71
Air ionization, 85–86
Air purifier, inspection of, 84–86
Amphiphiles, 27–30
amphiphilic molecules, 28–29
definition of, 27
emulsifiers, 28
role in biogenics, 29
Archipelago de Colon, *See* Galápagos Islands
Argentina, 16–19
Arkansas creation trial, 11–14
Artifacts
biologists view of, 155–56
dermatologists use of, 157
social science view of, 155
Atmosphere, 38
Atomic theory, 220–21
Ayatollah and Khayyam, 104–7

Bacteria, behavior of, 117–21
do they think? 121
genetically engineered, 117–19
patenting of, 117–21
Beetles, 31–35
Biochemistry, 197–200
Bio-energetics, 214–17
adenosine triphosphate (ATP), 214–17
Biosphere, 37–39
Bird excrement, 45–48
Blakemore, Richard, 209
Boiling water (in Ulysses), 227–31
Boltzmann, Ludwig, 13, 176
Book of Garlic by Lloyd Harris, 61
Boscovich, Rudjer, 174–77
Brave New World by Aldous Huxley, 141, 144–45
Bremerman, Hans, article "Complexity and Transcomputability" in *The Encylcopedia of Ignorance*, 114
Bryan, William Jennings, 7

Calf's Foot Jelly recipe, 62
Carnegie Museum in Pittsburgh, 18–19

INDEX

Cat's Cradle by Kurt Vonnegut, 202
Causality, 21
Celestial Mechanics by Pierre-Simon Laplace, 233
Cheating by student, 150
Churches, time of building, 99–103
Clausius, Rudolf, 219
Colorado
  extraction of minerals, 137–40
  legislation on land use, 137
  use vs. abuse of resources, 140
Columbus, Christopher, 182–85
  biography: *Admiral of the Ocean Sea* by Samuel Morison, 183
  poem by Joaquin Miller, 184
  record of voyages, 185
  relation to NASA, 183–84
Crapsey, Adelaide, quote from, 147
Creationists, 7–10
Creation-science, definition of, 12
Combustion engine invention, 146
Commitment in religion
  comparison to that in science and scientists, 110–12
  interviews with students, 108–10
Communicating science, 232
Composers, listeners rating, 126–29
Computers and modern life, 113–16
Court trials, Arkansas creationists, 11–14

Darrow, Clarence, 7
Darwin, Charles, 4, 33
  Descent of Man, 23
  Origin of the Species, 4, 7–9
  Voyage of the Beagle, 4, 17
Dental care
  of author, 55–58
  of John E. Mack, 51–58
Division barriers, 29–30
DNA, RNA, 40, 43–44, 162, 164, 190, 200
Dubrovnik conference, 174–77

Ecological sequence, 90–92
Ecosystems in our bodies, 89–92
Egyptian civilization
  contract with immortality, 101
  preoccupation with death, 100–101
Empedocles, 36–39
  leap into Mount Etna, 36
  teaching on make-up of matter, 36
Emulsifiers of oil and water, 27–29; *See also* Amphiphiles
Encyclopedia of Ignorance by Hans Bremermann, 114
Energy, 214–17
  electrical, 203–4
  entropy, 218–22
  flow in *Ulysses*, 227–31
Entropy, 218–22
  development of function, 221
  Rifkin, Jeremy, *Entropy: A New World View*, 221
  thermodynamics, 218
Escherich, Theodor, 190–93
  discovery of *E. coli* bacteria, 191
  study of children's diseases, 190–91
Escherichia coli (*E. coli*), 190–93
  discovery by Escherich, 191
  further studies by Twort and d'Herelle, 192
Extrasensory perception (ESP), 122–25

Feyerabend, Paul, *Against Method*, 111
First Three Minutes by Steven Weinberg, 22
Fitness of the Environment by Lawrence Henderson, 43
Flying machines
  crossing English channel, 170
  Gossamer Albatross, 170–73
Food and Drug Administration (FDA), 84–87

criteria for testing, 86–87
Devices Division, 84, 87
regulatory powers, 87
Fossil research by Hatcher, 17–19
Foucault, J. B. L., 166–69
pendulum in Yale museum, 166–69
Frankel, Richard, 209
Frost, Robert, poem quoted, 15

Gaia Hypothesis by James Lovelock, 41
Galápagos Islands, 3–6, 33
Darwin's comments on, 4–6
visit by author, 3–6
wonderment and joy of, 6
Gamwow, George, 161–65
Garlic
cure for atherosclerosis, 60–61
medical journal opinions, 60
therapeutic benefits of, 60–61
Generalists among scientists, 162–63
Generation gap, 70
Geochemistry, 37–39
Gilbert, William, 210
Glaciers, 201–4
Glacier Bay, Alaska, 201, 204
life history of, 204
God and the Astronomers by Robert Jastrow, 22
Golgi, Camillo, 155
"Golgi complex or apparatus," 155–56
Gossamer Albatross, 170–73
Gosse, Philip Henry
Book: *Omphalos, an Attempt to Untie the Geological Knot*, 8–10
contacts with Darwin, 7–10
Grantsmanship vs. research, 16–19
Guano
bird excrement on Nauru, 45–48
Guano act by US Congress, 47

Hahnemann, Christian Samuel, 76
Haldane, J. B. S., 31
Harris, Lloyd J., *Book of Garlic*, 61
Hatcher, John Bell
fossil research in Argentina, 17
methods of money raising, 18
Health, positive program of, 74–75
Henderson, Lawrence, *The Fitness of the Environment*, 43
d'Herelle, Felix, 192
Heisenberg, Werner, 163
uncertainty principle, 115
Hicks, Edward, *The Peaceable Kingdom*, 3
Holistic medicine, Chinese, 110
Homeopathy, 76–79
tenets of, 76
vs. allopathy, 78
Honesty, choice for, 153
Hudson, G. Evelyn, 33
Huxley, Aldous, *Brave New World*, 141, 144–45
Huygens, Christian, 146
Hydrosphere, 38

Iatrogenic disease, 64
Ice, ice age, 202–3
India conference, 14–15
Insects, 31–35
Isolation for a year, 147–49

Jastrow, Robert, *God and the Astronomers*, 22
Joyce, James, *Ulysses*, 227–31
Jurisgenic disease, 64–67
cure of, 67
intensity of symptoms, 65
social worker influence, 66

Khayyam, Omar, commentary on life (poetry), 104–7
King's College Chapel at Cambridge University, 99–103

Kuhn, Thomas, *Structure of Scientific Revolution*, 111

Lasaga, A. C., 29
Laplace, Pierre-Simon, *Celestial Mechanics*, 233
Lawyers, *See* Jurisgenic disease
Lecithin, 27–28
Learning from students, 72–75
Leisure time, 142–43
Life, views of, 40–44
  global view, 40–41
Listener rating charts for composers, 126–29
Lithosphere, 38
Lodestones, *See* Magnets
Longevity
  modern occupation with, 102–3
  replaces immortality, 102
Lovelock, James, Gaia hypothesis, 41

MacCready, Paul B., Jr. 170–73
Mack, Judge John E., 51–54
McNulty, Wilbur, 206
Magnets, magnetism, 209–13
  Gilbert's discovery of earth's magnetism, 210
  in animals and birds, 212
  magnetic microorganisms, 209, 210
  National Magnet Laboratory, 209
Mathematical abstractions, 232-35
Matter, make-up of, 36–39
  Biosphere added, 36–38
  Empedocles theory: air, earth, water, fire, 36
  geosphere, lithosphere, hydrosphere and atmosphere, 37
Maxwell, James Clerk, 123
  "Maxwell's demon," 123–25
May, Robert, 31–32
Mayonnaise
  combination of oil, water, 27–30
  origin of the word, 24
Medical profession
  changing perception of medicine, 83
  nation, international representation needed, 82
  numbers of students increase while applications decline, 82
  provincial character of, 81–83
  school selection policies, 80–83
Medical treatment
  group vs. individual, 75
  no action as optimal treatment, 79
Melville, Herman, *Moby Dick*, 186–89
  Melville as biologist, 188–89
  whale classification, 187–88
Metabolism, 41
  Chart of Intermediary Metabolism, 198–99
  metabolic processes, 199
Microbial behavior, 117–21
  recognition of attractants and repellents, 120
  response to stimuli, 120
Microorganisms
  freedom from at birth, 90
  present in our bodies, 89–92
Moby Dick by Herman Melville, 186–89
Molecular biology, 40–44
Molecular disorder, 13
Molybdenum, 223–26
  combination with nitrogen, 223–26
Monkeys, apes, 205–8
"Monkey trial," 7
Morison, Samuel, *Admiral of the Ocean Sea* (biography of Columbus), 183
Moses, John B. and Cross, Wilbur, *Presidential Courage*, 53

INDEX

Mouse, electronic, 114
Mystic religions, 108–10

NASA's space flights, 182, 183–84
Nash Ogden, couplet quoted, 45
National Institute of Health, 161, 165
Natural Resources
  exploitation of, 137–40
  use vs. abuse, 140
Natural selection, 4–6
Nature, isolation in, 146–49
Nature magazine, 95
Nauru, 45–48
von Neumann, John, 163
Newton's second law of motion, 233
Niches of animal species, 32–35
Nineteen Eighty-Four
  book by George Orwell, 141–43
  class of, 141–45
Nitrogen, 223–26
  combined with molybdenum, 223–26
Noogenesis, noosphere, 23–24, 26
Nutrient cycles, 42

Objectivity compromised by commitment, 108–12
Older people
  attitudes toward, 69
  examples of accomplishments, 68
Omar Khayyam, verses of commentary on life, 104–7
Onsager, Lars, 13
Orwell, George, *Nineteen Eighty-Four*, 141–45

Pace of Life, 95–98
  Bornstein study, 95
  walker pace measured, 95–98
Paleontology research, 17–19
Paper
  masses in modern world, 144
  multiple copy practice, 144
Paradox of paradoxes, 108–12
  objectivity compromised by commitment: in religion, 108–10; in science, 110–12
Pasteur, Louis, 175
Patagonia, fossil research in, 17–19
Peabody Museum, Yale, 166–69
Peaceable Kingdom by Edward Hicks, 3
Phosphorous, phosphates, in bird excrement, 45–48
Photochemistry, 41–42
Photosynthesis, 41
Plagiarism
  by student, 150
  letter from teacher, 150–53
Playing card analogy, 162–65
Poker, as a means of money-raising for research, 17–19
Pollution, 135
Polychlorinated biphenyl (PCB), 206–8
Pope, Alexander, quote from, 222
Presidential Courage by John Moses and Wilbur Cross, 53
Preventive Medicine, 73
Primate centers, 205–8
  Oregon Regional Primate Center, 205
Princeton University, 17–19
Prochronic theory, 9–10
Psychic phenomenon, 122–25

Quality of work, commitment to, 55–58

Redi, Francesco, 175
Reflective thought, 23–25
Religion and science
  debates, 7–10
  in Hindu teachings, 14–15
  in Teilhard de Chardin, 20–26
Retirement age, 68–71

Index

Rifkin, Jeremy, *Entropy*: A New World View, 221
RNA, *See* DNA
Robot, scotch drinking, 116
Roosevelt, President Franklin Delano, 51–54
Rotation of the earth, 166–69

Science and religion, *See* Religion and Science
Sea lion on Galápagos, 4–6
  maternal attachment of, 5
  natural selection example, 6
Seurat, Georges Pierre, 178–81
Shannon, Claude E., "The Mathematical Theory of Communication," 178–80
Silicon Valley, California, 113
Skin of Our Teeth by Thornton Wilder, 202
Slobodkin, L. B., *Is History a Consequence of Evolution?*, 25
Snow, C. P., 154, 157
Socrates
  influence on Athens, 135–36
  modern need for, 136
  questions asked by, 135
  visit to his prison, 133–36
Sodium in diet, 73
Sounds of life, 147–49
  in seasons in nature, 147–49
Species of animals
  niche of individual species, 32–35
  number in relation to size (chart), 32
Stefan, Josef, 176
Structure of Scientific Revolution by Thomas Kuhn, 111
"Succession," in biota, 90–92
Suicide, 151–52
  of Socrates, 133–35
Supreme Court ruling on patenting of bacteria, 117
Szilard, Leo, 163

Talmud
  quotation from, 72
  study in Jerusalem, 109
Teilhard, Pierre de Chardin, 20–26
  advent of life, 22
  causality discussions, 21
  noogenesis, 23–24, 26
  Phenomenon of Man, 20, 21, 23, 26
  reflective thought, 23, 25
Tetrachlorodibenzo-p-dioxin (TCDD), 206–8
Thermal units, 230
Thermodynamics, 12–14, 218
  Clausius studies in, 219
Time in building churches, 99–103
Twort, Frederick, 192

Ulysses by James Joyce, 227–31
Utrecht Cathedral, 99–103

Vernadsky, V. I., *La Biosphere*, 37
Vonnegut, Kurt, *Cat's Cradle*, 202

Wagner, Eugene, 234
Walking pace measurements, 95–98
Weinberg, Steven, *The First Three Minutes*, 22
Wilder, Thornton, *Skin of Our Teeth*, 202
Wilson, President Woodrow, 53–54
Words, uses of, 154–57
Wordsworth, William, quote from, 99

Yale University Peabody Museum, 160–69